Basic Statistical Physics

Basic Statistical Physics

Nandita Rudra · P Rudra

Retired from University of Kalyani, India

World Scientific

NEW JERSEY · LONDON · SINGAPORE · BEIJING · SHANGHAI · HONG KONG · TAIPEI · CHENNAI

British Library Cataloguing-in-Publication Data
A catalogue record for this book is available from the British Library.

To

All Our Students

In our endeavour to introduce you to the works of
Maxwell, Boltzmann, Gibbs, Einstein, Landau, Onsager, Feynman
and many other masters,
we have learned much.

Thank you.

Preface

mandakaviyaśaḥprārthī
    gamisyāmyupahāsyatām
prāṁśu labhye phale
    lobhādudbahuriba vāmanaḥ
athabā kṛtabāgdvāre
    baṁśesmin pūrvasuribhiḥ
maṇau vajrasamūtkīrṇe
    sūtrasyebāsti me gatiḥ

Like a dwarf aiming at
    fruits atop a lofty tree,
I may be attracting scorn for
    aspiring to be a bard.
Or, maybe, achievements of my
    predecessors will make me
A thread through the eyes on
    the gems to make a wreath.

Kālidāsa: Raghuvaṁśam 1:3-4   English translation.

This book grew out of the two-semester course of lectures that we have been delivering for a number of years to the first year Masters level students at the University of Kalyani. This book can be used in final year Senior level undergraduate Course or a first year Graduate level Course.

The first four chapters, § 1–§ 4, are the backbone of the discipline of Statistical Physics. Here the basic concepts of the subject as well as the description of states, their time evolution and the methods of Statistical Physics have been introduced and developed. All throughout, connection of Statistical Physics to Thermodynamics has been emphasized.

In the next four chapters, § 5–§ 8 , Classical Statistical Physics and its application to Chemical Equilibria, Interacting Systems and Strong Electrolytes have been explained.

The next three chapters, § 9–§ 11 are concerned with Quantum Statistical Physics and its application to two topical subjects, *viz.* Bose-Einstein Condensate and Statistical Astrophysics.

The next chapter, § 12, is an introduction to the systematics of Phase Transition and Landau Theory for systems *not very close* to the phase-coexistense line. We have also included a condensed introduction of the modern theory of phase transition *very close* to the phase-coexistense line, but have omitted the Renormalization Group Technique of calculating the Critical Indices.

Though the book is mainly comcerned with equilibrium processes, in § 13 we have discussed the Linear Response Theory of Irreversible Processes and Fluctuation-Dissipation Theorem. Physics consists of both static and dynamic processes and the large Resonance and Relaxation community always use this Kubo formalism to analyze their data of dynamic processes away from equilibrium

The purely mathematical tools that we needed in our development and which do not form a component part of Statistical Physics proper, have been collected in the last chapter § 14 as Mathematical Appendix.

Finally, we confess that this book is **not** a *Treatise on Statistical Physics*. Were it such a Treatise it would have covered at least two very important topics: (i) a topological discussion on validity of the *Ergodic Hypothesis*; (ii) a graphical method of calculation of the Virial Coefficients of Real Gases. Another lacuna is the absence of the modern field of activity in *Mesoscopic Systems*.

We have omitted many steps in the deductions of results in the body of the text and have set the completion of the proof for some of them as the problems set for the students. Most of the materials in the book are for ordinary 3-dimensional materials. The formulations for the 2-dimensional thin films and 1-dimensional linear polymers have been set as the other members of the problem sets.

Here is a note about numbering of the equations, figures, postulates, tables and theorems. They are all numbered sequentially in separate sections of each chapter. And we have consistently used the Gaussian units.

In an old and established subject like Statistical Physics there are many books that have by now become *classics* and many books that are mainly concerned with some still developing topics. Our target readers are masters level or beginner level graduate students who have not yet decided what their chosen line of research would be. So we have included just a few references as typical specimens and some comprehensive review articles on the newer applications of Statistical Physics.

We are aware of Dr Johnson's 1766 statement: 'People have now-a-days got a strange opinion that everything should be taught by lectures. Now, I cannot see that lectures can do so much good as reading the books from which the lectures are taken.' We, however, firmly believe that a good teacher is essential to the students for admiring the beauty, elegance and power of the subject. This book is a suggestion to the teacher that the course followed here may be one of many ways of developing the subject.

Nandita Rudra & P.Rudra.
Kolkata, India,
2009.

Notations and Fundamental Constants

α^{-1}	Inverse Fine Structure Constant $= 137.035\ 999\ 679$
β	$\frac{1}{k_B T}$
$B(m, n)$	Beta Function of arguments m & n
γ	Gyromagnetic Ratio
γ_0	Fundamental cell in phase space
Γ	Phase Space, Statistical Weight
$\Gamma(z)$	Gamma Function of argument z
$\delta(x)$	Dirac delta function of x
ϵ_F	Fermi Energy
$\zeta(z)$	Riemannian zeta function of argument z
η	Order Parameter
μ	Chemical Potential
μ_B	Bohr Magneton, $\frac{e\hbar}{2m_e}\ =\ 9.274\ 009\ 15 \times 10^{-24}\ \mathrm{JT^{-1}}$
ρ	Statistical Distribution Function
$\hat{\rho}$	Density Matrix
σ	Stefan-Boltzmann Constant
	$\frac{\pi^2}{60}\frac{k_B^4}{\hbar^3 c^2}\ =\ 5.670\ 400 \times 10^{-8}\ \mathrm{Wm^{-2}K^{-4}}$
$\vec{\sigma}$	Pauli Matrix
$\sigma_{\mu,\nu}(\omega)$	Electrical Conductivity Tensor
$\phi_{\hat{B},\hat{A}}(t)$	Response Function
$\Phi_{\hat{B},\hat{A}}(t)$	Relaxation Function
$\chi_{\hat{B},\hat{A}}(\omega)$	Admittance, Generalized Susceptibility, Magnetic Susceptibility Tensor
χ_L	Landau Diamagnetic Susceptibility
χ_P	Pauli Paramagnetic Susceptibility

$\hat{\mathbf{a}}$ Unit vector along $\mathbf{a}$

$\left[\hat{A}, \hat{B}\right]$ Commutator of two quantum mechanical operators $\hat{A}$ & $\hat{B}$

$[A, B]_{\mathrm{PB}}$ Poisson bracket of two classical functions A & B

$\mathbf{B}$ Magnetic Field

c Speed of light in vacuum $= 2.997\ 924\ 58 \times 10^8$ ms^{-1}

D Diffusion Constant

e Elementary Charge $= 1.602\ 176\ 487 \times 10^{-19}$ C

E Internal Energy

F Helmholtz' Free Energy

G Gibbs' Potential

h Planck's Constant $= 6.626\ 068\ 96 \times 10^{-34}$ Js

$\hbar$ $\frac{h}{2\pi} = 1.054\ 571\ 628 \times 10^{-34}$ Js

H Enthalpy

$\hat{\mathcal{H}}$ Hamiltonian operator

$\Im(z)$ Imaginary Part of complex number z

k_B Boltzmann Constant $= 1.380\ 650\ 4 \times 10^{-23}$ JK^{-1}

$\mathcal{K}$ Thermal Conductivity

K_T Isothermal Compressibility

m_e Mass of Electron $= 9.109\ 382\ 15 \times 10^{-31}$ kg

m_p Mass of Proton $= 1.672\ 621\ 637 \times 10^{-27}$ kg

n Number Density

$\hat{\mathcal{N}}$ Number operator

N_A Avogadro Constant $= 6.022\ 141\ 79 \times 10^{23}$ mol^{-1}

$\langle \hat{\mathcal{O}} \rangle$ Ensemble Average of the Operator $\mathcal{O}$

$\overline{\hat{\mathcal{O}}}$ Time Average of the Operator $\hat{\mathcal{O}}$

p_F Fermi Momentum

P Pressure

R Molar Gas Constant $= 8.314\ 472$ J mol^{-1}K^{-1}

$\Re(z)$ Real Part of complex number z

T Statistical Temperature

Tr Trace Operator

V Volume

Z Partition Function

Z_{gc} Partition Function for Grandcanonical ensemble

Contents

Chapter 1

Basic Concepts

1.1 Introduction

Macroscopic bodies consist of a large number of constituent atoms and molecules. The exact mechanical description of such bodies is almost impossible to obtain as the initial positions and momenta of the constituents are not known. Apart from the lack of knowledge of the initial conditions there is a weightier objection to obtaining a dynamical solution. If Δt is the time required for a measurement to be done then the *uncertainty in the measurement of energy* $\Delta E \sim \frac{\hbar}{\Delta t}$. Again if N is the number of particles in the system then the *separation between 2 energy levels of the system* is $\delta E \sim e^{-aN}$ where a is a constant, and with the N for any macroscopic system $N \sim 6.022 \times 10^{23} = Avogadro\ Constant$, we have $\Delta E \gg \delta E$. Thus we cannot talk about a definite energy of the system and the very concept of *Hamiltonian dynamics* is meaningless for such macroscopic bodies.

Thermodynamics gives the description of a macrosystem in terms of a finite number of parameters like pressure, volume, temperature and obtains relationship between them without taking into account the dynamics of the system. In Kinetic Theory of gases, however, we try to explain thermodynamic concepts like pressure, temperature *etc.* in terms of the dynamics of the gas molecules. Statistical Physics is the generalization of thermodynamics and kinetic theory, applicable to all systems and not only to gases. It obtains the thermodynamic quantities taking into account the interactions between the large number of atoms and molecules. This is done under the **thermodynamic limit** $\lim N \to \infty$ for the number of particles N of the system and $\lim V \to \infty$ for the volume V of the system so that number density $\frac{N}{V} = finite$. Generally the value of a double limit depends on the order in which the limits are taken as is shown in the following example.

$$\lim_{N \to \infty} \lim_{V \to \infty} \frac{NV}{N + V^2} = 0, \quad \text{while} \qquad (1.1.1)$$

$$\lim_{V \to \infty} \lim_{N \to \infty} \frac{NV}{N + V^2} = \infty. \qquad (1.1.2)$$

The restriction that $\frac{N}{V} = $ finite makes the limit finite and unique,

$$\lim_{\substack{N \to \infty, \\ V \to \infty, \\ \frac{N}{V} = \text{finite}}} \frac{NV}{N + V^2} = \frac{N}{V}, \qquad (1.1.3)$$

as any physical quantity should be.

We shall mention here that modern investigations have extended the domain of Statistical Physics also to **mesoscopic systems** where this limit breaks down.

The *largeness of the number* of constituents and the *complexities of their motion* give rise to some new laws, the statistical laws, on which the whole of equilibrium statistical physics is built. These laws can be enunciated as *two basic postulates*.

Postulate 1.1.1   **Hypothesis of Equal á-priori probability**: All microstates for an isolated system in equilibrium has equal probability of occupation.

Postulate 1.1.2   **Ergodic hypothesis**: The time average of a physical quantity which is the observed value of the quantity, is equal to the **ensemble average** of the quantity, which we derive from the theories of statistical physics.

A number of terms like **microstate** and **ensemble average** have been used in these postulates, which we shall define later.

Like building up any theory, in Statistical Physics also we start with the postulates, which are *intuitively obvious assumptions* that cannot be proved but may sometimes be justified. However, we must realize that the statistical mechanical description of macrosystems is probabilistic. This is due to the lack of knowledge of the initial coordinates and momenta and *not* due to any quantum mechanical uncertainties associated with measurements.

1.2 Master Equation and Hypothesis of Equal á-priori Probability

The hypothesis of **equal á-priori probability** can be justified from the following consideration. We consider a nearly isolated system approaching equilibrium by interacting with its surrounding to attain a final state of higher probability. The rate of change of population in each microstate of the syatem is governed by assuming the intuitively acceptable **Master Equation** connecting time rate of change of population, p_m, in the state m with the transition rates, $W_{k \to l}$, from the state k to the state l.

$$\frac{dp_m}{dt} = -\sum_{n \neq m} W_{m \to n} p_m + \sum_{n \neq m} W_{n \to m} p_n, \quad \text{separately for every m.} \quad (1.2.1)$$

The first term on the right hand side of Equation 1.2.1 denotes the rate of change of p_m due to transition *to* the state n, while the second term corresponds to transition *from* the state n to the state m. These transitions

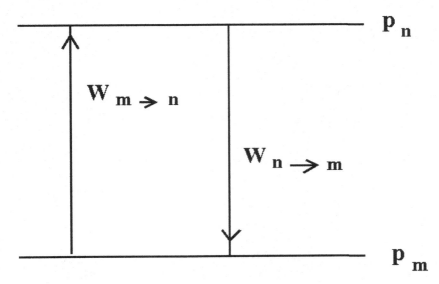

Fig. 1.2.1   Schematic diagram showing transitions between the states m and n.

are shown in Figure 1.2.1: Remembering that $W_{m \to m} = 0$ we can write

$$\frac{dp_m}{dt} = -\sum_{n} W_{m \to n} p_m + \sum_{n} W_{n \to m} p_n, \quad \text{separately for every m.} \quad (1.2.2)$$

Since in both Classical Mechanics as well as Quantum Mechanics $W_{n \to m} = W_{m \to n} = W_{(n,m)}$, depending only on the pair (n, m), we can write the **Master Equation** as

$$\frac{dp_m}{dt} = \sum_n W_{(n,m)} (p_n - p_m), \quad \text{separately for every m.} \qquad (1.2.3)$$

Since at statistical equilibrium $\frac{dp_m}{dt} = 0$ for every m, we immediately arrive at $p_m = p_n$ for all pairs of (n,m) as the **sufficient condition** of statistical equilibrium. Since the probabilities are proportional to the populations, we conclude that the **hypothesis of equal á-priori probability** is a **sufficient condition** for statistical equilibrium. It is also a **neccessary** condition if the transition rates $W_{(n,m)}$ satisfy certain conditions. We shall have an idea of these conditions as we consider the example of the 3-level system in § 1.2.1.

1.2.1 *Example of 3 Level Systems*

In this subsection we consider a 3-level system, depicted in Figure 1.2.2. The explicit forms of the Master Equation for this system at equilibrium are;

$$\frac{dp_1}{dt} \equiv W_{(1,2)} (p_2 - p_1) + W_{(1,3)} (p_3 - p_1) = 0, \qquad (1.2.4)$$

$$\frac{dp_2}{dt} \equiv W_{(1,2)} (p_1 - p_2) + W_{(2,3)} (p_3 - p_2) = 0, \qquad (1.2.5)$$

$$\frac{dp_3}{dt} \equiv W_{(1,3)} (p_1 - p_3) + W_{(2,3)} (p_2 - p_3) = 0. \qquad (1.2.6)$$

If the transition rates are all arbitrary, then $p_1 = p_2 = p_3$ i.e. **equal á-priori probability** will be the **neccessary** condition for statistical equilibrium. If, however, one of the levels (say, level # 3) is completely disconnected (i.e. $W_{(1,3)} = W_{(2,3)} = 0$) from all the other levels (i.e. level # 1 and level # 2) then the only conclusion we can draw from statistical equilibrium is that $p_1 = p_2$. **Hypothesis of equal á-priori probability** is thus a **neccessary condition** for statistical equilibrium only for connected states. This is actually true for any statistical system.

1.3 **Phase Space, Phase Point, Phase Trajectory**

To describe classically the states of a dynamical system one introduces the concept of **Phase Space**, which is a mathematical space consisting

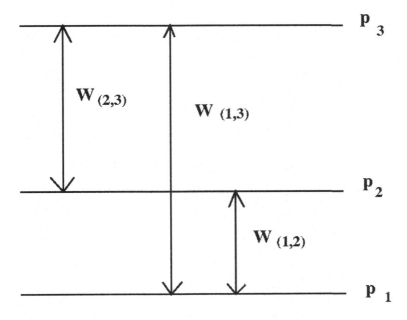

Fig. 1.2.2 Schematic diagram of the 3-level system.

of the *generalized coordinates* $q_1, q_2, \cdots, q_f$ and the *generalized momenta* $p_1, p_2, \cdots, p_f$, f being the *degree of freedom* of the system under study. The phase space will be denoted by the symbol Γ. It is evident that this $2f$- dimensional phase space has the *physical dimension* of f-dimensional action of Hamiltonian mechanics. Any point in the phase space will have fixed values of the f coordinates and the f momenta. This is called a **Phase Point**. At any point of time the state of the dynamical system is represented by a phase point in the phase space. As the system evolves in time the phase point traces out a curve in the phase space, called the **Phase Trajectory**. In Figure 1.3.1 we plot a schematic diagram of a phase trajectory in a phase space. The phase trajectories are *not self-intersecting*, because time evolution of a system has to be *unique*.

1.4 Statistical Distribution Function and Ergodic Hypothesis

We consider an elementary phasespace volume $\Delta\Gamma = \prod_{i=1}^{f} (\Delta q_i \Delta p_i)$ so that the i-th component of generalized coordinate lies between q_i and $q_i + \Delta q_i$ and

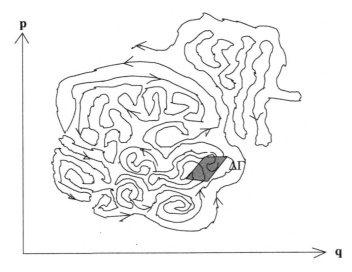

Fig. 1.3.1   Schematic representation of Phase Trajectory and infinitesimal Phase volume.

the i-th component of generalized momentum lies between p_i and $p_i + \Delta p_i$. Then the probability $\Delta\omega$ that the state of the system stays within $\Delta\Gamma$ can be defined in terms of time Δt spent by the phase trajectary inside $\Delta\Gamma$ as follows

$$\Delta\omega = \lim_{T\to\infty} \frac{\Delta t}{T}. \qquad (1.4.1)$$

Then the time average of any physical qantity $F(q, p, t)$ is given by

$$\overline{F} = \lim_{T\to\infty} \frac{1}{T} \int_{t_0}^{t_0+T} F(q, p, t)dt. \qquad (1.4.2)$$

By q and p we denote the generalized coordinates and the generalized momenta in a generic sense. We also tacitly assume that the value of the integral is independent of the starting point t_0.

We now define **statistical distribution function** as

$$d\omega = \rho(q, p, t)d\Gamma. \qquad (1.4.3)$$

The **statistical distribution function** $\rho(q, p, t)$ then represents the probability density function of the states of the system lying in the phase space region $d\Gamma$ and so satisfies the *normalization condition*

$$\int \rho(q,p,t)d\Gamma = 1. \qquad (1.4.4)$$

The statistical average

$$\langle F \rangle = \int F(q,p,t)\rho(q,p,t)d\Gamma \qquad (1.4.5)$$

can be looked upon as if there are many copies of the system having states distributed in the phase space at a particular instant of time according to the distribution function $\rho(q,p,t)$. This collection of identically prepared copies of the system is called a **Statistical Ensemble** and the statistical average is called **Ensemble Average**.

Having introduced time average and ensemble average of a physical quantity we arrive at the equality of the two averages

$$\lim_{T\to\infty} \frac{1}{T}\int_0^T F(q,p,t)dt = \int F(q,p,t)\rho(q,p,t)\,d\Gamma. \qquad (1.4.6)$$

The hypothesis can be proved for many special cases but a general *proof* is still lacking and in spite of many attempts of *proving* the Ergodic Hypothesis it remains a *Hypothesis*. However, it has been proved that if the phase space is *metrically connected* then the Ergodic Hypothesis is valid.

1.5 Statistical Fluctuation and Statistical Independence

We first note that though the experimentally measured quantities are equal to their ensemble averages, they are *not* absoutely constant; in smaller time intervals they fluctuate about these average values. We define **Statistical Fluctuation** as

$$\Delta F = F - \overline{F} \qquad (1.5.1)$$

with

$$\overline{\Delta F} = 0, \qquad (1.5.2)$$

Root Mean Square Fluctuation as

$$(\Delta F)_{\text{rms}} = \sqrt{\overline{(\Delta F)^2}} = \sqrt{\overline{(F - \overline{F})^2}}, \qquad (1.5.3)$$

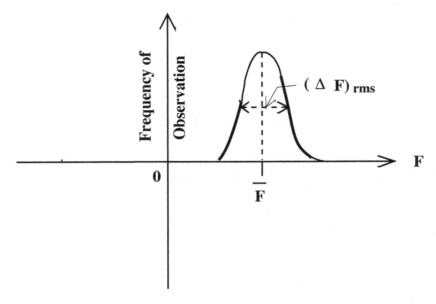

Fig. 1.5.1   Schematic plot of the frequency distribution of a physical quantity F, its average value $\overline{F}$ and its root mean square fluctuation $(\Delta F)_{\text{rms}}$.

and **Relative Fluctuation** as

$$(\Delta F)_{\text{rel}} = \frac{(\Delta F)_{\text{rms}}}{\overline{F}}. \tag{1.5.4}$$

In Figure 1.5.1 we sketch a typical frequency distribution of a physical quantity F, its average value $\overline{F}$ and its root mean square fluctuation $(\Delta F)_{\text{rms}}$.

The form of the statistical distribution function and fluctuation of a physical quantity takes particularly simple form for a special type of system. We consider a large isolated system which does *not* interact with its surrounding. We imagine the system to be made up of smaller parts called **subsystems** which also are *macroscopic*. These subsystems are *not* closed but interact with other subsystems surrounding them through their surfaces. Thus the interaction energy of a particular subsystem with others is much smaller (being $\propto R^2$ where R is its linear dimension) than the internal or volume energy $\propto R^3$. Thus the subsystems can be considered to be quasi-closed if they are observed for a time interval small compared to their relaxation time. It must, however, be noted that if observed for a long time these interaction energies cannot be neglected. As a matter of fact it is these interactions that wll lead to establishment of statistical equilibrium.

In this case the subsystems are called **Statistically Independent**. For statistically independent subsystems the state of one subsystem does *not* depend on the states of other subsystems. For systems consisting of N statistically independent subsystems the statistical distribution function of the total system is written in terms of those of the subsystems as

$$\rho(q,p,t) \stackrel{\text{def}}{=} \prod_{a=1}^{N} \rho_{(a)}\left(q^{(a)}, p^{(a)}, t\right). \tag{1.5.5}$$

If the subsystems are *not* statistically independent then such a factorization of the distribution function is *not* possible. Of course, for all systems the phase space volume can be factorized

$$d\Gamma = \prod_{a=1}^{N} d\Gamma_{(a)}. \tag{1.5.6}$$

Those physical quantities like number of particles, energy, entropy and other thermodynamic potentials which are sum of those of the constituent subsystems are called *additive*. For such additive physical quantities fluctuations of the total system can be written in terms of those of the constituent subsystems in particularly convenient form. If

$$F(q,p,t) = \sum_{a=1}^{N} F_{(a)}\left(q^{(a)}, p^{(a)}, t\right), \tag{1.5.7}$$

then, because of Equation 1.5.5

$$\overline{F} = \sum_{a=1}^{N} \overline{F_{(a)}} = Nm \tag{1.5.8}$$

and

$$\overline{(\Delta F)^2} = \overline{(F - \overline{F})^2}$$

$$= \overline{\left[\sum_{a=1}^{N} \left(F_{(a)} - \overline{F_{(a)}}\right)\right]^2}$$

$$= \sum_{a=1}^{N} \sum_{b=1}^{N} \left(F_{(a)} - \overline{F_{(a)}}\right)\left(F_{(b)} - \overline{F_{(b)}}\right)$$

$$= \sum_{a=1}^{N} \overline{\left(F_{(a)} - \overline{F_{(a)}}\right)^2} + \sum_{a=1}^{N}\sum_{b=1_{a\neq b}}^{N} \overline{\left(F_{(a)} - \overline{F_{(a)}}\right)\left(F_{(b)} - \overline{F_{(b)}}\right)}$$

$$= \sum_{a=1}^{N} \overline{\left(\Delta F_{(a)}\right)^2} + \sum_{a=1}^{N}\sum_{b=1_{a\neq b}}^{N} \overline{\left(\Delta F_{(a)}\right)\left(\Delta F_{(b)}\right)}$$

$$= \sum_{a=1}^{N} \overline{\left(\Delta F_{(a)}\right)^2} + \sum_{a=1}^{N}\sum_{b=1_{a\neq b}}^{N} \overline{\left(\Delta F_{(a)}\right)} \cdot \overline{\left(\Delta F_{(b)}\right)}$$

$$= \sum_{a=1}^{N} \overline{\left(\Delta F_{(a)}\right)^2} = N\sigma^2. \tag{1.5.9}$$

Because of the identity of the subsystems the mean m and the variance σ^2 have been taken as the same for each subsystem. The relative fluctuation of additive physical quantities for a system consisting of statistically independent subsystems is thus

$$(\Delta F)_{\text{rel}} = \frac{\sigma}{\sqrt{N}m} \propto \frac{1}{\sqrt{N}}. \tag{1.5.10}$$

Thus for macroscopic bodies when $N \sim 6.022 \times 10^{23}$ the relative fluctuation tends to zero and the average value of a physical quantity becomes the most probable value. This justifies our use of the statistical averages for the description of a macroscopic body. Any fluctuation from these average values are *extremely rare*. It should be mentioned that for **mesoscopic systems** of modern **nano-materials** the above thermodynamic limit cannot be applied and other techniques have to be used for their analysis.

A concrete physical example of this general result is manifest if we consider a system of N particles in a volume V. We imagine a given region of volume v within V. Let us imagine that n number of particles are in v. Because of statistical fluctuation n will fluctuate and we calculate $(\Delta n)_{\text{rel}}$. We assume that p is the probability of finding a particle in v so that $\frac{v}{V}$ is a very good physical estimate of p. The probability of finding n particle in v is thus the **binomial distribution** $W_B(n, N; p)$ given in § (14.7.2.1) of Mathematical Appendix. Using Equation 14.7.2 and Equation 14.7.3 of Mathematical Appendix, we arrive at the expected result

$$(\Delta n)_{\text{rel}} = \frac{1}{\sqrt{N}}\sqrt{\frac{1-p}{p}}. \tag{1.5.11}$$

1.6   Statistical Fluctuation and Generalized Susceptibility

Statistical fluctuations not only determine the applicability of the methods of statistical physics, it actually gives expressions for different response functions like compressibility, specific heat and thermal conductivity. These response functions control how the system reacts to application of external stimuli like mechanical pressure, thermal energy source and temperature gradient to it. The generic name for these response functions is **generalized susceptibility**. The name has been suggested from response of magnetic systems to external magnetic field.

We consider isothermal compressibility as a prototype of such a generalized susceptibility. As a result of fluctuation of volume V of the system, the number density n of particles changes as a random stochastic process. If ΔV and Δn are these changes and ΔF denotes the resulting change in the free energy, then the general principle states that the probability of volume change by this amount is

$$w(\Delta V) \propto \exp\left(-\frac{\Delta F}{k_B T}\right), \tag{1.6.1}$$

where k_B is the Boltzmann constant. Since F is minimum at equilibrium, so up to the first non-vanishing term

$$w(\Delta V) \propto \exp\left[-\left(\frac{\partial^2 F}{\partial V^2}\right)_T \frac{(\Delta V)^2}{2k_B T}\right]. \tag{1.6.2}$$

This is a Gaussian distribution in ΔV with mean zero and variance (*vide* § (14.7.2.3) in the Mathematical Appendix)

$$\overline{(\Delta V)^2} = \frac{k_B T}{\left(\frac{\partial^2 F}{\partial V^2}\right)_T}. \tag{1.6.3}$$

Using the thermodynamic relationships for pressure $P = -\left(\frac{\partial F}{\partial V}\right)_T$ and the isothermal compressibility $K_T = -\frac{1}{V}\left(\frac{\partial V}{\partial P}\right)_{T,N}$ we arrive at

$$\overline{(\Delta V)^2} = k_B T \overline{V} K_T. \tag{1.6.4}$$

We now calculate the fluctuation in the mean number density $\overline{n} = \frac{N}{\overline{V}}$ in two ways: first as a fluctuation in volume

$$\overline{(\Delta n)^2} = \left(\frac{\overline{n}}{\overline{V}}\right)^2 \overline{(\Delta V)^2} = \overline{n}^2 \frac{k_B T K_T}{\overline{V}}, \tag{1.6.5}$$

and then as a fluctuation in the total number of particles

$$\frac{\overline{(\Delta N)^2}}{\overline{N}^2} = \frac{\overline{(\Delta n)^2}}{\overline{n}^2} = \frac{k_B T K_T}{\overline{V}}. \tag{1.6.6}$$

Einstein showed that if electromagnetic radiation of intensity I_{inc} is incident on a medium in which the number density fluctuates then the incident radiation deviates from the path of the geometrical optics and gets scattered with intensity I_{sc} given by

$$\frac{I_{\text{sc}}}{I_{\text{inc}}} \propto \sqrt{\frac{\overline{(\Delta n)^2}}{\overline{n}^2}}. \tag{1.6.7}$$

The density fluctuation is thus directly measurable by experiments.

The most striking phenomenon of **Critical Opalescence** can be explained as a manifestation of number density fluctuation. When light is scattered by a *real gas* near its critical point, scattering increases enormously. The whole scattering chamber is then filled up with radiation giving a semblance of the surface of the naturally occurring mineral Opal, which is of dazzling milky white texture. This is the origin of the name of the phenomenon. A real gas consisting of interacting particles has $\left(\frac{\partial P}{\partial V}\right)_T \to 0$ and thus $K_T \to \infty$ as $T \to T_c$ the critical temperature. Thus the scattered intensity $I_{\text{sc}} \to \infty$ as T approaches the critical temperature T_c.

As we have mentioned connection between Number fluctuation and isothermal compressibility has been taken as a prototype. Similar connection exists for other response functions and fluctuations in corresponding physical variables. Connection between magnetic susceptibility χ and fluctuation of magnetic moment M

$$k_B T \chi = \overline{(\Delta M)^2} \tag{1.6.8}$$

and the **Einstein relation** between heat capacity C_v and fluctuation of energy E

$$k_B T^2 C_v = \overline{(\Delta E)^2} \tag{1.6.9}$$

are two other often-mentioned relations.

1.7 Generalized Ornstein-Zernicke Relation

Ornstein and Zernicke obtained expression for fluctuation of the number density in terms of inter-particle interaction. This gives a microscopic description of statistical fluctuation. We start with the **two-body density-correlation function**

$$n^{(2)}\left(\mathbf{r},\mathbf{r}'\right) \overset{\text{def}}{=} n^{(1)}\left(\mathbf{r}\right) n^{(1)}\left(\mathbf{r}'\right) g\left(\mathbf{r},\mathbf{r}'\right). \qquad (1.7.1)$$

The inter-particle interaction is incorporated in the inter-particle distribution function $g\left(\mathbf{r},\mathbf{r}'\right)$ and a sketch of this function is plotted in Figure 1.7.1. For isotropic systems $g\left(\mathbf{r},\mathbf{r}'\right) = g\left(R\right)$ where $\mathbf{R} = \mathbf{r} - \mathbf{r}'$ and $|\mathbf{R}| = R$. For

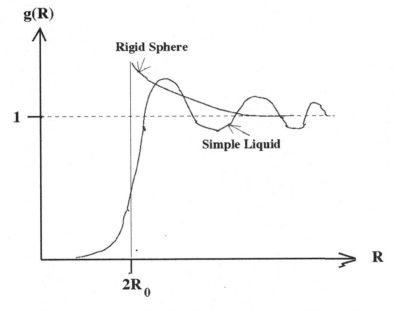

Fig. 1.7.1   A sketch of the inter-particle distribution function $g\left(R\right)$ is plotted for hard spheres of radii R_0 and simple liquids with valence electrons in the s and p shells.

non-interacting point particles $g(R) = 1.0$ for all values of R. Obviously as $R \to \infty$, $g(R) \to 1$. For systems consisting of rigid spheres of radii R_0, $g(R) = 0$ when $R < 2R_0$. For simple liquids with valence electrons in the s and p shells $g(R)$ are obtained from X-ray scattering data just above the melting temperature. Again, $n^{(1)}\left(\mathbf{r}\right)$ and $n^{(1)}\left(\mathbf{r}'\right)$ are different because of the existence of fluctuation in the medium.

With obvious mathematical operations we obtain

$$\overline{n^{(1)}(\mathbf{r})} = \overline{n} = \frac{\overline{N}}{\overline{V}}, \tag{1.7.2}$$

$$\overline{n^{(2)}(\mathbf{r},\mathbf{r}')} = \overline{n}^2 g(R), \tag{1.7.3}$$

$$\int\int\left[\overline{n^{(2)}(\mathbf{r},\mathbf{r}')} - \overline{n}^2\right]d^3rd^3r' = \overline{n}^2\int\int[g(R)-1]\,d^3rd^3r'$$
$$= \overline{n}^2V\int[g(R)-1]\,d^3\mathbf{R}. \tag{1.7.4}$$

We now express the number density and the two-body density correlation function in terms of Dirac Delta functions (§ 14.2 of the Mathematical Appendix)

$$n^{(1)}(\mathbf{r}) = \sum_j \delta(\mathbf{r_j} - \mathbf{r}), \tag{1.7.5}$$

$$n^{(2)}(\mathbf{r},\mathbf{r}') = \sum_{j\neq k} \delta(\mathbf{r_j} - \mathbf{r})\,\delta(\mathbf{r_k} - \mathbf{r}'), \tag{1.7.6}$$

$$\overline{n^{(1)}(\mathbf{r})} = \overline{\sum_j \delta(\mathbf{r_j} - \mathbf{r})}, \tag{1.7.7}$$

$$\overline{n^{(2)}(\mathbf{r},\mathbf{r}')} = \overline{\sum_{j\neq k} \delta(\mathbf{r_j} - \mathbf{r})\,\delta(\mathbf{r_k} - \mathbf{r}')}$$
$$= \overline{\sum_{j,k} \delta(\mathbf{r_j} - \mathbf{r})\,\delta(\mathbf{r_k} - \mathbf{r}')} - \overline{\sum_j \delta(\mathbf{r_j} - \mathbf{r})\,\delta(\mathbf{r_j} - \mathbf{r}')}$$
$$= \overline{n^2} - \delta(\mathbf{r} - \mathbf{r}')\overline{\sum_j \delta(\mathbf{r_j} - \mathbf{r})} = \overline{n^2} - \overline{n}\,\delta(\mathbf{r} - \mathbf{r}'). \tag{1.7.8}$$

We can now calculate

$$\int\int\left[\overline{n^{(2)}(\mathbf{r}-\mathbf{r}')} - \overline{n}^2\right]d^3rd^3r' = \int\int\left[\overline{n^2} - \overline{n}\delta(\mathbf{r}-\mathbf{r}') - \overline{n}^2\right]d^3rd^3r'$$
$$= \overline{N^2} - \overline{N} - \overline{N}^2. \tag{1.7.9}$$

Comparing Equation 1.7.4 and Equation 1.7.9 we obtain for the fluctuation

$$\frac{\overline{(\Delta N)^2}}{\overline{N}^2} = \frac{\overline{(\Delta n)^2}}{\overline{n}^2} = \frac{1}{\overline{N}}\left(1 + \overline{n}\int[g(R)-1]\,d^3\mathbf{R}\right) \tag{1.7.10}$$

and the **Ornstein-Zernicke Relation** for isothermal compressibility

$$K_T = \frac{1}{k_BT}\left(\frac{1}{\overline{n}} + \int[g(R)-1]\,d^3\mathbf{R}\right). \tag{1.7.11}$$

For a **perfect gas** of non-interacting point particles $g(R) = 1$ and we arrive at the well-known relation $K_T = \frac{1}{\bar{n}k_B T}$.

We shall later see in § 7.1 that the expression $\int [g(R) - 1] \, d^3\mathbf{R}$ appears also in the second virial coefficient of *real gases*.

1.8 Problems

Problem 1.1. Write down the Mathematical Condition that 'none of the levels in Figure 1.2.2 is isolated'. Prove that in this case at equilibrium $p_1 = p_2 = p_3$.

Problem 1.2. Draw the phase trajectory of a particle of mass m executing a one dimensional simple harmonic motion of frequency ω.

Problem 1.3. Prove Equation 1.5.9.

Problem 1.4. Prove Equation 1.5.11.

Problem 1.5. For a real gas satisfying the Virial Equation of state $P = \frac{\bar{N}k_B T}{V} \left[1 + B(T) \frac{N}{V} \right]$ obtain the expression for $\frac{\overline{(\Delta N)^2}}{\overline{N}^2}$.

Problem 1.6. Calculate the isothermal compressibility K_T, when the inter-particle distribution function is given by

$$g(R) = \begin{cases} 0, & \text{for } 0 \leq R \leq 2R_0 \\ g_0 \exp(-k_0 R) & \text{for } 2R_0 \leq R \leq \infty. \end{cases} \tag{1.8.1}$$

Chapter 2

Motion of Systems in Phase Space

2.1 Integral Invariants

In Classical Mechanics we know that every dynamical system with degrees of freedom f and described by a Hamiltonian $\mathcal{H}$ satisfying

$$\frac{dq_i}{dt} = \frac{\partial \mathcal{H}}{\partial p_i} = \Theta_i(q, p, t), \quad i = 1, \cdots, f \tag{2.1.1}$$

$$\frac{dp_i}{dt} = -\frac{\partial \mathcal{H}}{\partial q_i} = \Pi_i(q, p, t), \quad i = 1, \cdots, f \tag{2.1.2}$$

has a number of **invariants of motion** or **constants of motion**. These invariants are functions of the *generalized coordinates* q_i and *momenta* p_i. Poincaré has introduced certain quantities called **integral invariants** which are integrals over a p-dimensional region of the phase space. As the points in the p-dimensional region evolves in time, they will occupy a new p-dimensional region of the phase space at a later time. If an integral taken over this new region has the same value as the original one, then the integral is called an **Integral Invariant of order p**. For *incompressible fluids*, the integral which represents the volume of the fluid occupied initially in any given region is an integral invariant.

We state some important properties of integral invariants.

(i) The conditions for *integral invariancy* of
$\int \sum_{i=1}^{f} (M_i(q, p, t) \, \delta q_i + N_i(q, p, t) \, \delta p_i)$ are for all i

$$\frac{\partial M_i}{\partial t} + \sum_{j=1}^{f} \left(\frac{\partial M_i}{\partial q_j} \Theta_j + \frac{\partial M_i}{\partial p_j} \Pi_j + M_j \frac{\partial \Theta_j}{\partial q_i} + N_j \frac{\partial \Pi_j}{\partial q_i} \right) = 0, \tag{2.1.3}$$

$$\frac{\partial N_i}{\partial t} + \sum_{j=1}^{f} \left(\frac{\partial N_i}{\partial q_j} \Theta_j + \frac{\partial N_i}{\partial p_j} \Pi_j + M_j \frac{\partial \Theta_j}{\partial p_i} + N_j \frac{\partial \Pi_j}{\partial p_i} \right) = 0. \tag{2.1.4}$$

(ii) The *neccessary* and *sufficient* condition that
$\int \sum_{i=1}^{f} \left(\frac{\partial F(q,p,t)}{\partial q_i} \delta q_i + \frac{\partial F(q,p,t)}{\partial p_i} \delta p_i \right)$
is an integral invariant, is that $F(q,p,t)$ is an integral of motion.

(iii) If $x_i(q,p,t) = c_i$, $i = 1, \cdots, f$ and $y_i(q,p,t) = d_i$, $i = 1, \cdots, f$ are $2f$
numbers of integrals of motion then $\int \sum_{i=1}^{f} (M_i(x,y)\delta x_i + N_i(x,y)\delta y_i)$
with arbitrary functions M_i and N_i is an integral invariant of order 1.
Thus a Hamiltonian system possesses an *infinite number* of invariant
integrals.

We have not yet mentioned the nature of the phase space region chosen for
evaluating these integrals. If the phase space region is a *closed manifold*
then the integral invariant is called a **relative invariant integral** and for
the general case it is called an **absolute integral invariant**.

In Statistical Physics we shall study motion of a closed region of phase
space over time, as is depicted in Figure 2.1.1. Thus these are *relative*

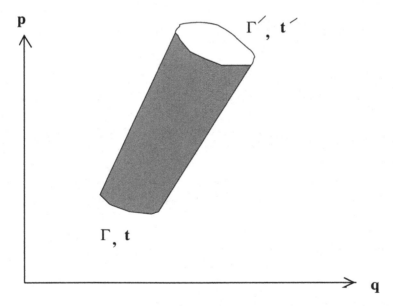

Fig. 2.1.1   Change of an elementary Phase volume over time.

integral invariants of order 2f of the form

$$I(t) = \int_{\Gamma \ at \ t} F(q,p)d\Gamma = \int_{\Gamma' \ at \ t'} F(q',p')d\Gamma' = I(t') \qquad (2.1.5)$$

so that $\frac{dI(t)}{dt} = 0$. As we have mentioned there are an infinite number of such integral invariants and we shall mainly use only a few of them.

One such is the **Action Integral**

$$J = \int_\Gamma dpdq = \int pdq. \qquad (2.1.6)$$

As it becomes clear, the fact that Action Integral is an invariant of motion was known even before the creation of Quantum Mechanics, which simply introduced its quantized values. It was also known to Classical physicists that Action Integral must have a *non-zero minimum value* so that no singularity appears in statistical physics. In classical statistical physics this elementary phase volume was assigned the notation γ_0. In quantum statistical physics γ_0 was assigned the value $(2\pi\hbar)^f$ for a system with number of degrees of freedom equaling f. In Figure 2.1.2 we have graphically plotted the change of the action integral for a free particle moving in 1 coordinate space dimension. From the geometrical fact that the areas of the two

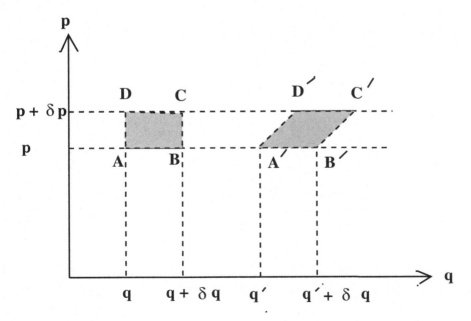

Fig. 2.1.2 Example of the change of an elementary Phase volume over time for a free particle moving in 1 coordinate dimension.

hatched regions are equal, equality of the **action integral** in this particular case follows.

2.2 Classical Liouville's Equation

If we want to obtain the Statistical Distribution Function, $\rho(q, p, t)$, we have to solve the differential equation for it. In order to obtain this differential equation, we shall invoke the results of § 2.1 as applied to the relative integral invariant $\int_\Gamma \rho(q, p, t) d\Gamma$. This integral physically denotes the *number of phase points* in the chosen phase volume, and hence must remain constant as the system evolves in time. We consider any arbitrary phase volume $d\Gamma = \prod_i dq_i dp_i$. As is evident from Figure 2.2.1 rate of change of phase point number in this phase volume due to drift along the q_i axis is

$$\left(\frac{d}{dt}\right)_{\text{drift along } q_i} \int \rho d\Gamma = -\int \left((\rho\dot{q}_i)_{q_i+dq_i} - (\rho\dot{q}_i)_{q_i}\right) \prod_k dp_k \prod_{k\neq i} dq_k$$

$$= -\int \frac{\partial}{\partial q_i}(\rho\dot{q}_i) d\Gamma.$$

Taking contribution to the drift term from all the generalized coordinates and momenta, we get the total contribution to drift as

$$\left(\frac{d}{dt}\right)_{\text{drift}} \int \rho d\Gamma = -\int \sum_i \left(\frac{\partial}{\partial q_i}(\rho\dot{q}_i) + \frac{\partial}{\partial p_i}(\rho\dot{p}_i)\right) d\Gamma \quad (2.2.1)$$

$$= -\int \sum_i \left(\frac{\partial\rho}{\partial q_i}\frac{\partial\mathcal{H}}{\partial p_i} - \frac{\partial\rho}{\partial p_i}\frac{\partial\mathcal{H}}{\partial q_i}\right) d\Gamma \quad (2.2.2)$$

$$= -\int [\rho, \mathcal{H}]_{\text{PB}} d\Gamma. \quad (2.2.3)$$

Here $\mathcal{H}$ is the Hamiltonian of the System and $[A, B]_{\text{PB}}$ is the Poisson Bracket of A & B (which is invariant under contact transformations) and we have used Hamilton's canonical equations of motion

$$\dot{p}_i = -\frac{\partial\mathcal{H}}{\partial q_i}, \text{ and } \dot{q}_i = \frac{\partial\mathcal{H}}{\partial p_i}. \quad (2.2.4)$$

The intrinsic rate of change of the number of phase points in this phase volme is

$$\left(\frac{d}{dt}\right)_{\text{intrinsic}} \int \rho d\Gamma = \int \frac{\partial\rho}{\partial t} d\Gamma. \quad (2.2.5)$$

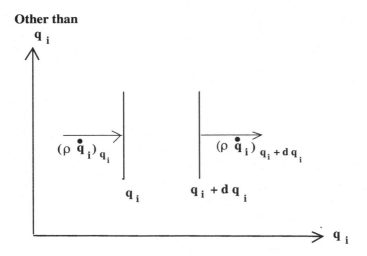

Other than

Fig. 2.2.1   Motion of Phasepoints across a hypersurface perpendicular to the q_i axis.

Equating the two rates and remembering that this is true for any arbitrary $d\Gamma$, we get the **Classical Liouville's Eqation**

$$\frac{\partial \rho}{\partial t} = [\mathcal{H}, \rho]_{\mathrm{PB}}. \qquad (2.2.6)$$

If we consider the **Hydrodynamic Equation of Motion** for the density, ρ, of an **Incompressible Fluid**

$$\frac{D\rho}{Dt} \equiv \frac{\partial \rho}{\partial t} + [\rho, \mathcal{H}]_{PB} = 0, \qquad (2.2.7)$$

where $\frac{D}{Dt}$ denotes the **hydrodynamic derivative**, we immediately conclude that the phase points behave like an *incompressible fluid* in the phase space. This also proves that $\int \rho \, d\Gamma$ over any phase volume is an *integral invariant*.

There is a quantum mechanical form of Equation 2.2.6 which we shall establish in § 2.5. However we can intuitively arrive at it, if we remember that in quantum mechanics every classical function is replaced by the coresponding operator and the classical Poisson bracket $[A, B]_{PB}$ is replaced by the quantum mechanical commutator through $(i\hbar)^{-1} \left[\hat{A}, \hat{B}\right]$.

2.3 Role of Energy

It is apparent from Equation 2.2.6 that, for stationary states, the statistical distribution function should be expressed in terms of those functions of the generalized coordinates and momenta q and p that remain constant as the subsystems evolve along the phase trajectory. These combinations of q and p are the *constants of motion*.

We can further restrict the constants of motion that will appear in the statistical distribution function ρ of the combined system. Since for **statistically independent** subsystems (denoted by the subscript a) $\rho = \prod_a \rho_a$ we conclude that $\ln \rho$ is an **additive function of the constants of motion**. From mechanics we know that there are only *seven* such constants: (i) energy E, (ii) linear momentum **P** and (iii) angular momentum **L**. We can thus write

$$\ln \rho_a \;=\; \ln A_a \;-\; \beta E_a \;+\; \vec{\alpha}\cdot\mathbf{P}_a \;+\; \vec{\gamma}\cdot\mathbf{L}_a, \qquad (2.3.1)$$

where the constants β, $\vec{\alpha}$ and $\vec{\gamma}$ are the same for all subsystems.

Thus in a frame of reference where the total linear momentum $\mathbf{P} = \sum_a \mathbf{P}_a$, and the total angular momentum $\mathbf{L} = \sum_a \mathbf{L}_a$ are zero, $\ln \rho_a$ and hence ρ itself depends only on the energy of the subsystem.

2.4 Quantum Mechanical Density Matrix

In Quantum Statistical Physics the different subsystems in an ensemble having the same quantum state are characterized by the same state but with different random phases and we need an extra ensemble average (denoted by an overbar) over and above the Quntum Mechanical Expectation value (denoted by angular bracket) of the dynamical variables $\hat{A}$. So in Quantum Statistical Physics, instead of the quantum mechanical expectation value given by Equation 14.8.6 in the Mathematical Appendix, we have the *Quantum Statistical* Expectation value

$$\overline{\langle \hat{A} \rangle} = \sum_{m,n} \overline{C_m(t)^* C_n(t)} A_{m,n} \qquad (2.4.1)$$

$$= \sum_{m,n} \rho_{n,m} A_{m,n} \qquad (2.4.2)$$

$$= Tr(\rho \hat{A}) \;=\; Tr(\hat{A}\rho), \qquad (2.4.3)$$

where

$$\rho_{n,m} = \overline{C_m(t)^* C_n(t)} \qquad (2.4.4)$$

is the **Quantum Mechanical Density Matrix** equivalent to the classical **statistical distribution function**.

The Trace operator, Tr, has the following properties:

$$\text{Additivity}: \quad Tr\left(\sum_a \hat{A}_a\right) = \sum_a Tr\,\hat{A}_a, \tag{2.4.5}$$

$$\text{Associativity}: \quad Tr\left(\hat{A}\left(\hat{B}\hat{C}\right)\right) = Tr\left(\left(\hat{A}\hat{B}\right)\hat{C}\right), \tag{2.4.6}$$

$$\text{Cyclic Permuation}: \quad Tr\left(\hat{A}_1\hat{A}_2\hat{A}_3\cdots\hat{A}_n\right) = Tr\left(\hat{A}_2\hat{A}_3\cdots\hat{A}_n\hat{A}_1\right). \tag{2.4.7}$$

It should be noted that Equation 2.4.7 is true only for finite dimensional matrices. The quantum mechanical commutator $[\hat{q},\hat{p}] = i\hbar\hat{I}$ is a classic example where Equation 2.4.7 is violated, since $\hat{q}$ and $\hat{p}$ are infinite dimánsional matrices.

The Density Matrix has the following properties:

$$\text{Normalization}: \quad Tr\,\hat{\rho} = 1, \tag{2.4.8}$$

$$\text{Hermiticity}: \quad \hat{\rho} = \hat{\rho}^{\dagger}. \tag{2.4.9}$$

In the Coordinate Representation

$$\hat{\rho}\left(\mathbf{r},\mathbf{r}'\right) = \sum_{m,n}\phi_m\left(\mathbf{r}'\right)^*\rho_{n,m}\phi_n\left(\mathbf{r}\right). \tag{2.4.10}$$

Since $\hat{\rho}$ is *Hermitean* and hence *Normal*, it is *diagonalizable by similarity transformation*:

$$\hat{\rho}^d = S\hat{\rho}S^{-1} = \begin{pmatrix} w_1 & 0 & 0 & \cdots \\ 0 & w_2 & 0 & \cdots \\ 0 & 0 & w_3 & \cdots \\ \vdots & \vdots & \vdots & \ddots \end{pmatrix}. \tag{2.4.11}$$

If we use Equations 2.4.9, 2.4.8 we get

$$0 \leq w_n \leq 1, \quad \text{for all n,} \tag{2.4.12}$$

$$\sum_n w_n = 1, \tag{2.4.13}$$

and thus

$$w_n^2 \leq w_n. \tag{2.4.14}$$

Thus

$$\hat{\rho}^2 \leq \hat{\rho}. \tag{2.4.15}$$

In this representation, $w_n = \rho_{n,n}(t) = \overline{|C_n(t)|^2}$ is the statistical probability of finding the system in the state $|n\rangle$, and the diagonalized matrix is of the form

$$\rho^d = \sum_n |n\rangle w_n \langle n|. \tag{2.4.16}$$

The **Density Matrix** $\hat{\rho}$ is thus the Quantum Mechanical equivalent of the Classical **Statistical Distribution Function** ρ.

A Statistical **mixed state** is one where the subsystems in an ensemble have *completely random phase factors*; and a **pure state** is one where all of them have the *same phase factor*. A purely quantum mechanical system is thus a statistical pure system. For a pure state the inequality Equation 2.4.15 reduces to strict equality.

$$\rho^2 = \rho. \tag{2.4.17}$$

2.5   Quantum Liouville's Equation

Like the Classical *Statistical Distribution Function*, ρ, the Quantum Mechanical *Density Matrix* $\hat{\rho}$ also satisfies a differential equation and in analogy to the classical case we call it the **Quantum Mechanical Liouville's Equation**.

In the Energy (E_n) Representation the $C_n(t)$s in Equation 14.8.4 are of the form

$$C_n(t) = e^{-iE_n t/\hbar} C_n, \tag{2.5.1}$$

$$i\hbar \frac{\partial}{\partial t} C_n(t) = E_n C_n(t). \tag{2.5.2}$$

The matrix element $\rho_{m,n}$ of the Density Matrix $\hat{\rho}$ satisfies:

$$i\hbar \frac{\partial}{\partial t} \rho_{m,n} = (E_m - E_n) \rho_{m,n}$$

$$= \left(\hat{\mathcal{H}}\hat{\rho}\right)_{m,n} - \left(\hat{\rho}\hat{\mathcal{H}}\right)_{m,n},$$

where

$$\hat{\mathcal{H}} = \text{Hamiltonian of the system.}$$

We thus arrive at the **Quantum Mechanical Liouville's Equation**:

$$i\hbar\frac{\partial}{\partial t}\hat{\rho} = \left[\hat{\mathcal{H}},\hat{\rho}\right]. \tag{2.5.3}$$

Here $[\cdots,\cdots]$ denotes the quantum mechanical commutator. If we compare it with Heisenberg's equation of motion for any operator, in particular for the density matrix $\hat{\rho}$

$$i\hbar\frac{d}{dt}\hat{\rho} = i\hbar\frac{\partial}{\partial t}\hat{\rho} + \left[\hat{\rho},\hat{\mathcal{H}}\right] \tag{2.5.4}$$

we immediately realize that Quantum Mechanical Liouville's equation asserts that $\hat{\rho}$ remains constant and the system in in equilibrium state.

If, moreover, the system is *stationary*, i.e.

$$\frac{\partial}{\partial t}\hat{\rho} = 0, \tag{2.5.5}$$

then

$$\left[\hat{\rho},\hat{\mathcal{H}}\right] = 0. \tag{2.5.6}$$

Thus $\hat{\rho}$ is a function of the integrals of motion of the system.

If, moreover, the subsystems are *statistically independent*, then

$$\hat{\rho} = \prod_a \hat{\rho}_a, \tag{2.5.7}$$

$$\ln\hat{\rho} = \sum_a \ln\hat{\rho}_a. \tag{2.5.8}$$

Thus $\ln\hat{\rho}$, being an additive function, is a *linear function of the basic additive integrals of motion* of the dynamical system: Number operator $\hat{\mathcal{N}}$, Hamiltonian $\hat{\mathcal{H}}$, linear momentum $\hat{\mathbf{P}}$ and angular momentum $\hat{\mathbf{L}}$.

$$\ln\hat{\rho}_a = -\beta\left(\hat{\mathcal{H}}_a - \mu\hat{\mathcal{N}}_a\right) + \vec{\alpha}\cdot\hat{\mathbf{P}}_a + \vec{\gamma}\cdot\hat{\mathbf{L}}_a. \tag{2.5.9}$$

If we remember Equation 2.4.8 and use a suitable guage, we get for such *stationary* and *Statistically independent* systems:

$$\hat{\rho} = e^{-\beta\left(\hat{\mathcal{H}} - \mu\hat{\mathcal{N}}\right)}/Tr\left(e^{-\beta\left(\hat{\mathcal{H}} - \mu\hat{\mathcal{N}}\right)}\right). \tag{2.5.10}$$

Here $\beta = \frac{1}{k_B T}$ and μ is the chemical potential.

For those systems where the number of particles of each subsystem is moreover constant, we get:

$$\hat{\rho} = e^{-\beta\hat{\mathcal{H}}}/Tr\left(e^{-\beta\hat{\mathcal{H}}}\right). \tag{2.5.11}$$

Here, again, $\beta = \frac{1}{k_B T}$.

We now calculate as an example the Density Matrix of a quantum mechanical system consisting of a single spin-$\frac{1}{2}$ paricle of gyromagnetic ratio γ placed in an external magnetic induction **B**. For this system the Hamiltonian is

$$\hat{\mathcal{H}} = -\frac{1}{2}\gamma\hbar\left(\vec{\sigma}\cdot\mathbf{B}\right). \tag{2.5.12}$$

Using Equation 2.5.11, and the properties of the Pauli Matrices given in Eqations 14.6.4 and 14.6.5 of the Mathematical Appendix we obtain for this system

$$\hat{\rho} = \frac{1}{2}\left[\hat{\mathbf{I}} + \tanh\left(\frac{1}{2}\beta\gamma\hbar B\right)\left(\vec{\sigma}\cdot\mathbf{B}/B\right)\right]. \tag{2.5.13}$$

2.6   Problems

Problem 2.1.   Prove Equations 2.1.3 and 2.1.4.

Problem 2.2.   Show that $p\delta q - q\delta p$ is an integral invariant for the dynamical system $\frac{dq}{dt} = \frac{p}{m}$, $\frac{dp}{dt} = -m\omega^2 q$.

Problem 2.3.   Calculate the Action Integral $J = \int dp\,dq$ for a particle of mass m executing an 1-dimensional simple harmonic motion $q = A\cos(\omega t)$.

Problem 2.4.   Prove Equations 2.4.5, 2.4.6 and 2.4.7.

Problem 2.5.   Prove Equation 2.5.13.

Problem 2.6.   Show that for finite β Equation 2.5.13 describes a system in *mixed state*.

Chapter 3

States in Statistical Physics

3.1 Microscopic and Macroscopic States

Development of statistical physics depends on the concept of **statistical states**. There are two types of states involved. **Microscopic** or **microstate** of a system is determined by the states of the constituent atoms or molecules whereas **macroscopic** or **macrostate** is determined by the equilibrium values of macroscopic variables like total energy, total magnetic moment *etc.*

As a text book example we look at a system of 3 spin-1/2 particles, each having magnetic moment μ, fixed in space. In presence of an external magnetic field of magnitude B, there will be different possible alignments of spins and magnetic moments. The spin projection quantum number can have two values $+\frac{1}{2}$ or $-\frac{1}{2}$ along *any* direction, say that along B. Magnetic moment of each particle parallel to B is $+\mu$ and antiparallel to B is $-\mu$. The energy of each particle in the magnetic field will thus be $-\mu B$ or $+\mu B$ depending on the orientation of the magnetic moment. We designate the particle state $(+)$ or $(-)$ for the projection of magnetic moment $+\mu$ or $-\mu$ respectively. We now tabulate in Table 3.1.1 the possible 3-particle states. The macrostates are defined by unique values of total energy or what is the same thing in this example the total magnetic moment. Here the states #1 and #8 each defines separate macrostates. All the states #2 — #4 constitute a single macrostate; and similar is the case for states #5 — #7. The indices in the first column of Table 3.1.1 indicate different microstates. It is evident from this example that each macrostate may contain different numbers of microstates.

Table 3.1.1   Different particle states and their energies for a 3-particle system.

State index	Magnetic moment alignment for particle			Total Magnetic Moment	Total Energy
	1	2	3		
1	+	+	+	$+3\mu$	$-3\mu B$
2	+	+	-	$+\mu$	$-\mu B$
3	+	-	+	$+\mu$	$-\mu B$
4	-	+	+	$+\mu$	$-\mu B$
5	+	-	-	$-\mu$	$+\mu B$
6	-	+	-	$-\mu$	$+\mu B$
7	-	-	+	$-\mu$	$+\mu B$
8	-	-	-	-3μ	$+3\mu B$

3.2   Statistical Weight and Density of States

Statistical Weight $\Gamma(E)$ for the macroscopic variable E, the energy, is defined as the number of microstates with energy $\leq E$. Classically, for a system with degree of freedom equal to f any set of *allowed* values for the generalized coordinates $(q_1, q_2, \cdots, q_f)$ and the geeralized momenta $(p_1, p_2, \cdots, p_f)$ will define a *microstate* and we calculate the *allowed volume* of the phase space where the energy $\leq E$. We now take an elementary *cell* γ_0 in the phase space to which a *single microstate* can be assigned. If we now divide the *allowed volume* by γ_0 we obtain the **statistical weight**. In *classical statistical physics* no concrete value could be assigned to γ_0. In *quantum statistical physics*, however, we can assign definite value to γ_0. Bohr-Sommerfeld quantization rule assigns the value $2\pi\hbar$ to the *mechanical action* for motion in each dimension. So for a system with degree of freedom f, γ_0 will have the value $(2\pi\hbar)^f$. But even long before the advent of quantum mechanics classical physicists knew that a minimum value has to be assigned to γ_0 in order to avoid unsurmountable difficulties. We have used the *same* symbol both for the phase space and the statistical weight so that we do *not* proliferate the number of symbols.

Density of States is now defined in terms of the Statistical Weight as the *number of microstates per unit energy interval*

$$\Omega(E) = \frac{\partial \Gamma(E)}{\partial E}. \tag{3.2.1}$$

3.3   Examples: Non-interacting One- and N-Particle Systems and Spin-1/2 Particles

We now evaluate staistical weights and density of states for three different physical systems that we shall later use very fequently.

(i) *A single free particle of mass m moving in a 3-dimensional space V*:
The *Hamiltonian* for this system is

$$\mathcal{H} = \frac{1}{2m} \left(p_x^2 + p_y^2 + p_z^2 \right) = E. \tag{3.3.1}$$

Using $p = \sqrt{2mE}$, where E is the energy, we get

$\Gamma(E) = (\text{Number of microstates with Energy (macrostate)} \leq E)$

$$= \frac{V \times \text{Volume of sphere in momentum space with radius} \sqrt{2mE}}{(2\pi\hbar)^3}$$

$$= \frac{\frac{4}{3}\pi V}{(2\pi\hbar)^3} (2m)^{\frac{3}{2}} E^{\frac{3}{2}}, \tag{3.3.2}$$

$$\Omega(E) = \frac{2\pi V}{(2\pi\hbar)^3} (2m)^{\frac{3}{2}} E^{\frac{1}{2}}. \tag{3.3.3}$$

(ii) *A system of N non-interacting free particles of mass m moving in a 3-dimensional space V*:
The *Hamiltonian* for the system is

$$\mathcal{H} = \frac{1}{2m} \sum_{j=1}^{N} \left(p_{xj}^2 + p_{yj}^2 + p_{zj}^2 \right). \tag{3.3.4}$$

We again write for energy E

$$E = \frac{p^2}{2m} = \frac{1}{2m} \sum_{j=1}^{N} \left(p_{xj}^2 + p_{yj}^2 + p_{zj}^2 \right). \tag{3.3.5}$$

In order to calculate $\Gamma(E)$ we need the volume of the $3N$-dimensional sphere of radius $p = \sqrt{2mE}$ in the $3N$-dimensional momentum space and we use Equation 14.13.1 of § 14.13 in Mathematical Appendix. For calculating the statistical weight a distinction is made whether the particles are **distinguishable** or **identical**.

For a system of *distinguishable* particles we have the statistical weight

$$\Gamma_{\text{dist}} = \frac{\pi^{\frac{3N}{2}}}{\Gamma\left(\frac{3N}{2} + 1\right)} \frac{(2m)^{\frac{3N}{2}} V^N}{(2\pi\hbar)^{3N}} E^{\frac{3N}{2}}, \tag{3.3.6}$$

and the density of state

$$\Omega_{\text{dist}} = \frac{\pi^{\frac{3N}{2}}}{\Gamma\left(\frac{3N}{2}\right)} \frac{(2m)^{\frac{3N}{2}} V^N}{(2\pi\hbar)^{3N}} E^{\frac{3N}{2} - 1}. \tag{3.3.7}$$

Unlike in the case of *distinguishable* particles, if we make a *permutation* of the generalized coordinates and momenta of the N *identical* particles we obtain a new phase point in the phase space, but this state corresponds to the old state with the N *identical* particles permuted and is *not* physically distinct. Since there are $N!$ distinct permutations of the N particles, the $N!$ *microstates* thus obtained are *not* different. Thus the number of *microstates* of a system consisting of N *identical* particles is obtained by dividing the corresponding number of *microstates* of *distinguishable* particles by $N!$. We thus have the statistical weight

$$\Gamma_{\text{ident}} = \frac{\pi^{\frac{3N}{2}}}{\Gamma\left(\frac{3N}{2} + 1\right)} \frac{(2m)^{\frac{3N}{2}} V^N}{(2\pi\hbar)^{3N} N!} E^{\frac{3N}{2}}, \tag{3.3.8}$$

and the density of state

$$\Omega_{\text{ident}} = \frac{\pi^{\frac{3N}{2}}}{\Gamma\left(\frac{3N}{2}\right)} \frac{(2m)^{\frac{3N}{2}} V^N}{(2\pi\hbar)^{3N} N!} E^{\frac{3N}{2} - 1}. \tag{3.3.9}$$

This type of problem regarding identical particles arose in classical thermodynamics in connection with diffusion of gases. When gas molecules in two separated chambers, each of volume V and containing N molecules, were allowed to mix by merging the two enclosures, entropy of the total system increases by $2N \log 2$. If we consider the molecules as *identical*, then our physical intuition tells us that there should not be any change in the value of entropy. This is the famous **Gibbs' paradox**. Gibbs gave the resolution of this paradox which is essentially what has been stated here. Though the overwhelming majority of physicists are happy with Gibbs' formulation, we shall mention that there are other explanations, notably the one by von Neumann and another by Information theorists.

(iii) *System of N non-interacting spin-$\frac{1}{2}$ particles each of magnetic moment*
μ *in external magnetic field* **B**:
If the magnetic moment of a particle is parallel to **B** then its energy
is $-\mu B$, and if it is anti-parallel to **B** then the corresponding energy
is $+\mu B$. If n of the particles have their magnetic moments parallel to
B then this *macrostate* has the energy

$$E = n(-\mu B) + (N - n)(+\mu B) = (N - 2n)\mu B. \qquad (3.3.10)$$

We then have

$$n = \frac{N}{2} - \frac{E}{2\mu B}. \qquad (3.3.11)$$

Let p be the probability that a particle has its magnetic moment μ
parallel to **B**. Later on we shall see how to obtain p. But for our
purpose the value of p is *not* necessary. Only thing that we need here
is that there is a *fixed p*. The number of way we can choose n particles
out of a total of N particles is thus the number of *microstates* that form
the above *macrostate*, *i.e.* the *statistical weight*. From § (14.7.2.1) for
this binomial distribution of the Mathematical Appendix we then find

$$\Delta\Gamma(E) = \frac{N!}{n!\,(N-n)!} = \frac{N!}{\left(\frac{N}{2} - \frac{E}{2\mu B}\right)!\left(\frac{N}{2} + \frac{E}{2\mu B}\right)!}. \qquad (3.3.12)$$

This 2-level system is a convenient example to explain many concepts
of statistical physics and we shall use it very often in future.

3.4 Entropy and Boltzmann's Principle

In classical thermodynamics Boltzmann enunciated a famous relation

$$S = k_B \log W, \qquad (3.4.1)$$

where W was called **thermodynamic probability**. But in thermodynamics this *thermodynamic probability* was never very convincingly defined. In
Statistical physics we are now in a position to give a precise definition of
this *thermodynamic probability* in terms of *statistical weight* and **density
of states**. We now state the **Boltzmann's Principle** defining **thermodynamic entropy** S as

$$S(E) = k_B \log \Omega(E). \qquad (3.4.2)$$

In the thermodynamic limit when the number of constituents of the system
$N \to \infty$ an equivalent definition is

$$S(E) = k_B \log \Delta\Gamma(E). \qquad (3.4.3)$$

In order that $S(E)$ thus defined is really the thermodynamic entropy,
this $S(E)$ must satisfy the two important properties of thermodynamic
entropy:

(i) S(E) is an additive function for systems having non-interacting subsystems.
(ii) S(E) assumes its maximum value as the system attains equilibrium.

The first property follows easily when we remember that the statistical
weight can be factorized in terms of the statistical weights of its *statistically
independent* subsystems and logarithm of a product is sum of logarithms of
the factors. The second property is true because in equilibrium the system
attains a state whose statistical weight is a maximum.

We have already emphasized that in statistical physics everything is
expressed in terms of either the classical *statistical distribution function*
ρ, or the *quantum mechanical density matrix* $\hat{\rho}$. So we shall now proceed
to express S in terms of ρ. We consider a system with N non-interacting
identical subsystems. Such systems are called *statistical ensembles* which
we shall study in details in § 4. Each member of this ensemble has energies
$E_1, E_2, \cdots$. Let N_1 of the subsystems have energy E_1, N_2 of them have
energy E_2 and so on. Evidently we have

$$N = N_1 + N_2 + \cdots \qquad (3.4.4)$$

$$E = N_1 E_1 + N_2 E_2 + \cdots . \qquad (3.4.5)$$

Using the *multinomial distribution* which is a generalization of *binomial
distribution* (§ (14.7.2.1) of Mathematical Appendix) the number of ways
of distributing the N subsystems in groups $N_1, N_2, \cdots$ is given by $\frac{N!}{N_1!N_2!\cdots}$.
The statistical weight is thus

$$\Delta\Gamma(E) = \sum_{\substack{N_1 + N_2 + \cdots = N \\ N_1 E_1 + N_2 E_2 \cdots = E}} \frac{N!}{N_1!N_2!\cdots}. \qquad (3.4.6)$$

One of the summands corresponding to the *most probable distribution*
$(\overline{N_1}, \overline{N_2}, \cdots)$ has a very large value compared to all the other terms. So we

can approximate to a very large accuracy

$$\Delta\Gamma = \frac{N!}{N_1! N_2! \cdots}. \tag{3.4.7}$$

Thus the entropy of the system consisting of N identical subsystems is

$$S_N = k_B \log \frac{N!}{N_1! N_2! \cdots} = -N k_B \sum_n w_n \log w_n, \tag{3.4.8}$$

and the entropy of each subsystem is

$$S = -k_B \sum_n w_n \log w_n = -k_B Tr \left(\hat{\rho} \log \hat{\rho} \right). \tag{3.4.9}$$

Equation § 3.4.9 is thus the statistical definition of thermodynamic entropy. In arriving at this relation we have used the Stirling's Theorem 14.11.1 of Mathematical Appendix and the expression $w_n = \frac{N_n}{N}$ for the n-th eigenvalue of the density matrix (*vide* Equation 2.4.11).

As an example we shall now calculate the entropy of N non-interacting spin-$\frac{1}{2}$ particles for which the statistical weight has already been obtained

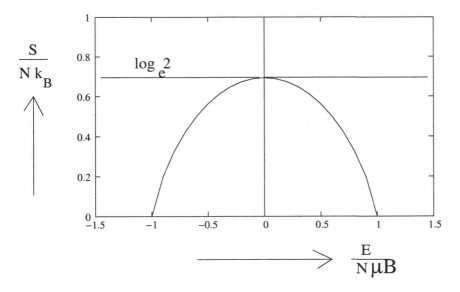

Fig. 3.4.1   Plot of Entropy S in units of the Boltzmann's constant k_B as a function of Energy E for a system of N noninteracting spin-$\frac{1}{2}$ particles of magnetic moment μ in external magnetic field **B**.

in Equation 3.3.12. Using that expression for the statistical weight and Boltzmann's Principle we obtain the entropy

$$S = Nk_B \log 2 - \frac{Nk_B}{2}\left(1 + \frac{E}{N\mu B}\right)\log\left(1 + \frac{E}{N\mu B}\right)$$
$$-\frac{Nk_B}{2}\left(1 - \frac{E}{N\mu B}\right)\log\left(1 - \frac{E}{N\mu B}\right) \qquad (3.4.10)$$

and we have sketched it in Figure 3.4.1.

3.5   Boltzmann's H-Theorem

When Maxwell presented his celebrated velocity distribution law every physicist instinctively realized that is *the* law, though no one was quite satisfied with Maxwell's 'proof'. Maxwell himself was aware of this shortcoming and he himself more than once tried to correct the shortcomings. However, it is only after Boltzmann showed that Maxwell's distribution law follows from the *H-* **theorem** that physics community was ultimately satisfied. This theorem reappeared in Information theory in the name of entropy. The non-essential difference was that physicists used the transcendental number e as the base of logarithm while the information theorists used the prime number 2 as the base. **Boltzmann's H-function** is defined as

$$H \stackrel{\text{def}}{=} \sum_n p_n \log p_n, \qquad (3.5.1)$$

where p_n is the probability of occupation of the n-th state. On comparing with Equation 3.4.9 we see that Boltzmann's H-function is the negative of the system's entropy measured in units of the Boltzmann's constant k_B. Using Master Equation 1.2.3 we immediately obtain

$$\frac{dH}{dt} = \sum_{m,n} W_{(m,n)}(p_m - p_n)(\log p_n - 1)$$
$$= \frac{1}{2}\sum_{m,n} W_{(m,n)}(p_m - p_n)(\log p_n - \log p_m). \qquad (3.5.2)$$

Since logarithm is a monotonous function of its argument and $W_{(n,m)} \geq 0$, we arrive at **Boltzmann's H-theorem**

$$\frac{dH}{dt} \leq 0. \qquad (3.5.3)$$

It can be shown that at equilibrium H-function is actually a minimum. This shows that the law of increase of entropy can be established from the dynamics of the system without recourse to the concept of any heat engine, once we assume the validity of the Master Equation.

Boltzmann established Maxwell's velocity distribution law by considering conservative binary collisions of the gas molecules at equilibrium.

3.6 Problems

Problem 3.1. Show that the statistical weight for a 1-dimensional simple harmonic oscillator of mass m and frequancy ω is $\Gamma(E) = \frac{E}{\hbar\omega}$.

Problem 3.2. Calculate the statistical weight for a system of N noninteracting electrons of mass m moving in a linear polymer of length L.

Problem 3.3. Calculate the statistical weight for a system of N noninteracting electrons of mass m moving in a thin film of area A.

Problem 3.4. Calculate the statistical weight for a system of N *identical particles* distributed among G energy eigenstates, so that each eigenstate can contain *no more than 1 particle*.

Problem 3.5. Calculate the statistical weight for a system of N *identical particles* distributed among G energy eigenstates *without any restriction on the occupancy of an eigenstate*.

Problem 3.6. Show that $\lim_{N\to\infty} k_B \log \Gamma(E) = \lim_{N\to\infty} k_B \log \Omega(E)$.

Problem 3.7. Prove Equation 3.4.9.

Problem 3.8. Prove Equation 3.4.10.

Problem 3.9. Prove Equation 3.5.2.

Problem 3.10. Prove that at equilibrium Boltzmann's H-function is a minimum.

Chapter 4

Statistical Ensembles

4.1   Introduction

Since statistical physics is probabilistic in nature, it is imperative to construct an assembly of a large number of identically prepared systems, each satisfying the same constraints (like energy and number conservation *etc.*) that the system under study satisfies. This assembly is called an **ensemble** and the probability of obtaining a specific event is given by the fraction of systems in the ensemble that are characterized by the occurrence of this particular event. Depending on how the subsystems in the ensemble interact among themselves the ensembles are given different names. Thus a **Microcanonical Ensemble** is for **isolated subsystems** when the subsystems exchange neither *energy* nor *particles* among themselves. In a **Canonical Ensemble** the subsystems can exchage only energy among themselves but *not* particles; whereas in a **Grandcanonical Ensemble** *both* energy and particles are exchanged among the subsystems.

4.2   Microcanonical Distribution Function

For an *isolated system* in equilibrium we construct the **microcanonical ensemble**. Each element of this ensemble satisfies the energy and the particle number conservation laws and the situation is shown pictorially in Figure 4.2.1. Thus if the isolated system has an energy lying between E_0 and $E_0 + \delta E$, then the probability

$$w_n = w = constant \text{ for } E_0 \leq E_n \leq E_0 + \delta E, \qquad (4.2.1)$$

as is shown in Figure 4.2.2. Going to the limit $\delta E \to 0$, we obtain

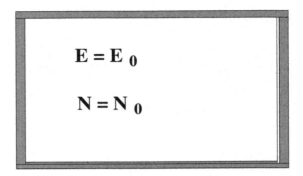

Fig. 4.2.1   Schematic depiction of a microcanonical ensemble with constant energy $E = E_0$ and number of particle $N = N_0$.

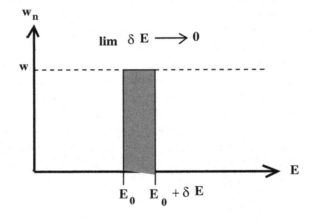

Fig. 4.2.2   Graphical depiction of the Microcanonical distribution.

$$dw = \text{Constant} \times \delta(E - E_0)d\Gamma \qquad (4.2.2)$$

$$= \text{Constant} \times \delta(E - E_0)e^{\frac{S(E)}{k_B}} dE. \qquad (4.2.3)$$

In arriving at the last expression we have used Boltzmann's Principle, Equation 3.4.2. Here $d\Gamma$ denotes the number of microstates with energy between E and $E + dE$.

If, moreover, we want to show the functional dependence on the number of particle, we write

$$dw = \text{Constant} \times \delta(E - E_0)\delta(N - N_0)d\Gamma \qquad (4.2.4)$$

$$= \text{Constant} \times \delta(E - E_0)\delta(N - N_0)e^{\frac{S(E,N)}{k_B}}\,dEdN. \qquad (4.2.5)$$

This is the basic distribution function and distribution functions for the other ensembles will be obtained from this basic one.

4.3 Canonical (Gibbs) Distribution Function

4.3.1 *Thermodynamic Temperature and Distribution Function*

The **canonical ensemble** is constructed for a system in contact with a heat bath. We consider the whole system $A_0(E_0, \Gamma_0)$ comprising of a subsystem $A(E, \Gamma)$ under study and the environment (heat bath) $A'(E', \Gamma')$ in which it is immersed, so that

$$E_0 = E + E', \quad \text{and} \quad d\Gamma_0 = d\Gamma d\Gamma'. \qquad (4.3.1)$$

The situation is shown in Figure 4.3.1 Since the whole system $A_0(E_0, \Gamma_0)$ is

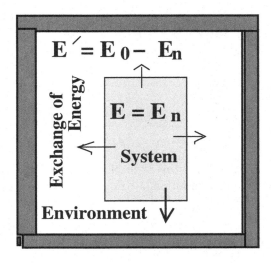

Fig. 4.3.1 Schematic sketch of a subsystem in a canonical ensemble exchanging only energy and *not* particle with the environment.

microcanonical, the probability that the subsystem is in a definite energy state E_n can be obtained by integrating the probability, Equation 4.2.3, of the microcanonical system over the states of the environment

$$
\begin{aligned}
w_n &= \text{Constant} \times \int e^{\frac{S'(E')}{k_B}} \delta\left(E_n + E' - E_0\right) dE', \\
&= \text{Constant} \times \exp\left[S\left(E_0 - E_n\right)/k_B\right], \\
&= Ae^{-\frac{E_n}{k_B T}}.
\end{aligned}
\tag{4.3.2}
$$

In the course of arriving at Equation 4.3.2 we have used the statistical definition of **Temperature**, T, as

$$
\frac{1}{T} \overset{\text{def}}{=} \left(\frac{\partial S}{\partial E}\right).
\tag{4.3.3}
$$

This statistical definition agrees with the thermodynamic definition. In equilibrium, entropy is a non-decreasing function of energy. Thus for a system in statistical equilibrium, its temperature is always positive. Since entropy as a function energy can be calculated from the states of the system, temperature can also be defined from the dynamics of the system. The constant A is determined from the normalization of probability $\sum_n w_n = 1$:

$$
\frac{1}{A} = \sum_n \exp\left(-\frac{E_n}{k_B T}\right).
\tag{4.3.4}
$$

The quantum mechanical density matrix for this canonical ensemble is

$$
\hat{\rho}_{\text{can}} = \frac{\exp\left(-\frac{\hat{\mathcal{H}}}{k_B T}\right)}{Tr\left(\exp\left(-\frac{\hat{\mathcal{H}}}{k_B T}\right)\right)}.
\tag{4.3.5}
$$

Since Gibbs extensively studied this type of ensemble it is also called the **Gibbs' Ensemble**. The distribution law Equation 4.3.2 for the Gibbs' ensemble is almost always referred to as Boltzmann's distribution.

4.3.2 *Spin-1/2 Particles and Negative Temperature*

Already in Figure 3.4.1 we have sketched the entropy of the two-level spin-$\frac{1}{2}$ system as a function of its energy. It is seen that when $E < 0$ its temperature is positive, indicating that the system is in statistical equilibrium. On the other hand, when $E > 0$ the temperature is negative and the system is in non-equilibrium state. The negative temperature non-equilibrium state is *energetically higher* than the positive temperature equilibrium state. Anticipating the later result of Equation 4.3.6 for the partition function, we conclude from the *finiteness* of the partition function that negative temperature can occur for systems with finite number of energy levels.

Non-equilibrium negative temperature states in magnetic systems were experimentally created by A.A.Abragam and his group. A magnetic system was placed in a constant magnetic field and the magnetic energy levels were populated according to Boltzmann law, Equation 4.3.2, the lower energy levels being more populated. The magnetic field is then suddenly reversed in direction. The level that was previously the energetically lowest level now becomes energetically the highest level but the spins do not get time to realign to the new equilibrium population among the levels; the new energetically upper levels have now more partcles than the new energetically lower levels. A negative temperature non-equilibrium state was thus created. The lifetime of this negative energy state is determined by the spin-lattice relaxation rate of the system, which was selected to be large.

4.3.3 Partition Function and Different Thermodynamic Functions

Normalization $\sum_n w_n = 1$ will allow us to obtain perhaps the most important function of statistical physics, the **Partition Function**

$$Z \stackrel{\text{def}}{=} \sum_n \exp(-E_n/k_B T) = \frac{1}{A}. \qquad (4.3.6)$$

The symbol has come from the first letter of its German name **Zuständsumme** literally meaning **Sum-over-states**, the term found in older English text books. As the name indicates it is a sum over *states*; so if some energy state is degenerate, that degeneracy has to be taken into account.

Most of the thermodynamic variables and functions can be expressed in terms of the partition functions, Z. For example the mean energy $\overline{E}$ is given by

$$\overline{E} = \sum_n w_n E_n = \frac{1}{Z} \sum_n E_n e^{-\beta E_n} = -\frac{\partial}{\partial \beta} \log Z, \qquad (4.3.7)$$

where $\beta = \frac{1}{k_B T}$.

Using the expression for w_n from Equation 4.3.2 and the definition, Equation 4.3.6, of the partition function, we can write the expression, Equation 3.4.9, for entropy as

$$S = -k_B \sum_n w_n \log w_n = -k_B \log A + \frac{\overline{E}}{T}, \qquad (4.3.8)$$

and obtain for Helmholtz' Free Energy

$$F = \overline{E} - TS = -k_B T \log Z. \qquad (4.3.9)$$

The macroscopic work (*not* a perfect differential) δW done *by* the system when an *external parameter* x changes by an infinitesimal amount dx is given by

$$\delta W = \sum_n w_n \left(-\frac{\partial E_n}{\partial x} \right) dx = \frac{1}{\beta Z} \left(\frac{\partial Z}{\partial x} \right) dx = k_B T \left(\frac{\partial}{\partial x} \log Z \right) dx. \qquad (4.3.10)$$

The corresponding **generalized force** X can be identified as

$$X = k_B T \left(\frac{\partial}{\partial x} \log Z \right). \qquad (4.3.11)$$

Thus if x is volume V, then the pressure is given by

$$P = k_B T \left(\frac{\partial}{\partial V} \log Z \right). \qquad (4.3.12)$$

4.3.4 *System of Linear Harmonic Oscillators in Canonical Ensemble*

We now discuss the important case of N independent linear harmonic oscillators in thermal equilibrium. This is a model for many physical systems. The Partition Function of the system as also its Free Energy can be expressed in terms of those of each oscillator of *characteristic frequency* ω_k for the k-th oscillator as follows.

$$Z = \prod_{k=1}^{N} Z_k, \quad \text{and} \quad F = \sum_{k=1}^{N} F_k = -k_B T \sum_{k=1}^{N} \log Z_k. \qquad (4.3.13)$$

It is far simpler to evaluate the quantities appearing in Equation 4.3.13 for the quantum mechanical system where the k-th oscillator with the *characteristic frequency* ω_k in the n-th quantum state has the energy $E_n^{(k)} = \hbar\omega_k(n + \frac{1}{2})$. So we obtain the partition function Z_k, the Free

Energy F_k and the mean energy $\overline{E_k}$ for the k-th oscillator

$$Z_k = \sum_{n=0}^{\infty} \exp\left[-\frac{\hbar\omega_k}{k_BT}\left(n+\frac{1}{2}\right)\right]$$

$$= \frac{\exp\left(-\frac{\hbar\omega_k}{2k_BT}\right)}{1-\exp\left(-\frac{\hbar\omega_k}{k_BT}\right)}, \tag{4.3.14}$$

$$F_k = -k_BT\log Z_k$$

$$= \frac{\hbar\omega_k}{2} + k_BT\log\left[1-\exp\left(-\frac{\hbar\omega_k}{k_BT}\right)\right], \tag{4.3.15}$$

$$\overline{E_k} = -\frac{\partial}{\partial\beta}\log Z_k$$

$$= \frac{\hbar\omega_k}{2} + \frac{\hbar\omega_k}{\exp\left(\frac{\hbar\omega_k}{k_BT}\right)-1}. \tag{4.3.16}$$

Hence for the system of N oscillators we have

$$F = \sum_{k=1}^{N}\left[\frac{\hbar\omega_k}{2} + k_BT\log\left(1-\exp\left(-\frac{\hbar\omega_k}{k_BT}\right)\right)\right], \tag{4.3.17}$$

$$\overline{E} = \sum_{k=1}^{N}\left[\frac{\hbar\omega_k}{2} + \frac{\hbar\omega_k}{\exp\left(\frac{\hbar\omega_k}{k_BT}\right)-1}\right]. \tag{4.3.18}$$

The classical result will follow if we take the limit $\hbar\to 0$.

4.3.5 *Energy Fluctuation in Canonical Ensemble and Equivalence of Canonical and Microcanonical Ensembles*

We now analyze fluctuation phenomenon for Canonical Ensembles. Using the notation $(\cdots)' = \frac{\partial}{\partial\beta}(\cdots)$, where $\beta = \frac{1}{k_BT}$ we have (*vide* Equation 4.3.7)

$$\overline{E} = -\frac{Z'}{Z}, \tag{4.3.19}$$

and

$$\overline{E^2} = \frac{1}{Z} \sum_n E_n^2 \exp\left(-\beta E_n\right)$$

$$= \frac{Z''}{Z}. \tag{4.3.20}$$

Mean square fluctuation in energy is thus

$$\overline{(\Delta E)^2} = \overline{E^2} - \overline{E}^2$$

$$= \frac{Z''}{Z} - \left(\frac{Z'}{Z}\right)^2$$

$$= \frac{\partial}{\partial\beta}\left(\frac{Z'}{Z}\right)$$

$$= -\frac{\partial\overline{E}}{\partial\beta}. \tag{4.3.21}$$

We have thus arrived at **Einstein's Relation**

$$\overline{(\Delta E)^2} = k_B T^2 \frac{\partial\overline{E}}{\partial T} = k_B T^2 C_v, \tag{4.3.22}$$

where the Heat Capacity at constant volume $C_v \propto N$. Since we also have $\overline{E} \propto N$ we get as a special case of the general result, Equation 1.5.10

$$(\Delta E)_{\text{rel}} \propto \frac{1}{\sqrt{N}}. \tag{4.3.23}$$

For macroscopic systems with $N \sim 10^{23}$ the fluctuations in energy of a canonical ensemble are so rare that we can take the energy of the canonical ensemble to be constant like in a microcanonical ensemble. In this sense canonical ensembles are said to be equivalent to microcanonical ensembles.

4.4   Grandcanonical Distribution Function

4.4.1   *Dependence of Thermodynamic Functions on Number of Particles*

Grandcanonical ensembles are constructed when both energy E and particle number N change. Since the classical thermodynamical relations were obtained for constant particle number, we first consider the changes to be introduced in the thermodynamical relations when the particle number N is *not* a constant.

For constant N the following thermodynamical relations are true.

$$
\begin{array}{lll}
\text{Enthalpy} & H & = E + PV, \\
\text{Helmholtz' Free Energy} & F & = E - TS, \\
\text{Gibbs' Potential} & G & = E - TS + PV, \\
\text{Second Law of Thetmodynamics } TdS & = dE + PdV.
\end{array}
\tag{4.4.1}
$$

The functional dependence of these functions on the variables S, T, P, V and their corresponding partial derivatives are

$$
\begin{array}{llll}
dE = TdS - PdV, & E \equiv E(S,V), & \left(\frac{\partial E}{\partial S}\right)_V = T & \left(\frac{\partial E}{\partial V}\right)_S = -P, \\
dH = TdS + VdP, & H \equiv H(S,P), & \left(\frac{\partial H}{\partial S}\right)_P = T & \left(\frac{\partial H}{\partial P}\right)_S = V, \\
dF = -SdT - PdV, & F \equiv F(T,V), & \left(\frac{\partial F}{\partial T}\right)_V = -S & \left(\frac{\partial F}{\partial V}\right)_T = -P, \\
dG = -SdT + VdP, & G \equiv G(T,P), & \left(\frac{\partial G}{\partial T}\right)_P = -S & \left(\frac{\partial G}{\partial P}\right)_T = V.
\end{array}
\tag{4.4.2}
$$

Of the four *independent variables* Entropy S, Pressure P, Temperature T and Volume V, only T and P are *intensive* variables while Entropy S and Volume V are extensive variables $\propto N$.

We can now introduce the particle number N and obtain the expressions of energy, enthalpy, Helmholtz' free energy and Gibbs' potential per particle

$$
\frac{E}{N} \equiv e\left(\frac{S}{N}, \frac{V}{N}\right),
\tag{4.4.3}
$$

$$
\frac{H}{N} \equiv h\left(\frac{S}{N}, P\right),
\tag{4.4.4}
$$

$$
\frac{F}{N} \equiv f\left(T, \frac{V}{N}\right),
\tag{4.4.5}
$$

$$
\frac{G}{N} \equiv \mu\left(T, P\right).
\tag{4.4.6}
$$

The **Chemical Potential** $\mu(T, P)$ is defined as

$$
\mu(T, P) = \frac{G}{N} = \left(\frac{\partial G}{\partial N}\right)_{P,T}.
\tag{4.4.7}
$$

Since $G \equiv G(T, P, N)$ we have

$$
dG = \left(\frac{\partial G}{\partial T}\right)_{P,N} dT + \left(\frac{\partial G}{\partial P}\right)_{T,N} dP + \left(\frac{\partial G}{\partial N}\right)_{T,P} dN
\tag{4.4.8}
$$

$$
= -SdT + VdP + \mu dN,
\tag{4.4.9}
$$

where we have used Equation 4.4.7 and Equation 4.4.2.

We can now calculate for the other thermodynamic functions

$$dF(T, V, N) = -SdT - PdV + \mu dN; \quad \left(\frac{\partial F}{\partial N}\right)_{T,V} = \mu, \quad (4.4.10)$$

$$dH(S, P, N) = TdS + VdP + \mu dN; \quad \left(\frac{\partial H}{\partial N}\right)_{S,P} = \mu, \quad (4.4.11)$$

$$dE(S, V, N) = TdS - PdV + \mu dN; \quad \left(\frac{\partial E}{\partial N}\right)_{S,V} = \mu. \quad (4.4.12)$$

The second law of thermodynamics for *variable number of particles* become

$$dS(E, V, N) = \frac{1}{T}dE + \frac{P}{T}dV - \frac{\mu}{T}dN, \quad (4.4.13)$$

with

$$\left(\frac{\partial S}{\partial E}\right)_{V,N} = \frac{1}{T}, \quad \left(\frac{\partial S}{\partial V}\right)_{E,N} = \frac{P}{T}, \quad \left(\frac{\partial S}{\partial N}\right)_{E,V} = -\frac{\mu}{T}. \quad (4.4.14)$$

For *many-component system* we have the relations

$$G = \sum_k G_k = \sum_k N_k \mu_k(T, P); \quad \mu_k = \left(\frac{\partial G_k}{\partial N_k}\right) = \left(\frac{\partial G}{\partial N_k}\right). \quad (4.4.15)$$

In the above calculations it is tacitly assumed that the system is a 3-dimensional one, which most of the systems of practical interest are. But there are two types of systems of physical importance that are of dimensions 2 and 1. These are the **thin films** and **linear polymers**. In the case of 2-dimensional thin films the second law of thermodynamics takes the form

$$TdS = dE + SdA, \quad (4.4.16)$$

where S is the **Surface Tension** playing the role of pressure P, and A is the **area** of the thin film system playing the role of volume V. In the case of 1-dimensional linear polymer the law is

$$TdS = dE + \mathcal{F}dL, \quad (4.4.17)$$

where $\mathcal{F}$ is the **Tension** in the string playing the role of pressure P and L is the **length** of the linear polymer playing the role of volume V.

4.4.2 *Chemical Potential and Distribution Function*

The subsystems in a **grandcanonical encemble** exchange both energy and particle among themselves, so that

$$E_{nN} + E' = E_0, \quad N + N' = N_0. \tag{4.4.18}$$

The energy-scheme of a N-particle subsystem depends on the number of particles of the subsystem. As an example we note that energy of a subsystem having a single linear harmonic oscillator depends on a single quantum number n, while that for a subsystem of two linear harmonic oscillators depends on two quantum numbers n_1 and n_2. So we have included the number of particles N in the symbol for energy of the subsystem. The primed quantities as before refer to the environment. The total system consisting of the subsystem and the environment having total energy E_0 and total number of particles N_0 form a microcanonical system. The situation is shown in Figure 4.4.1. A subsystem of a grandcanonical ensemble is also called an **open system**.

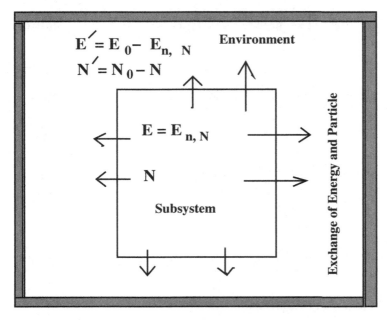

Fig. 4.4.1 Sketch of the subsystem in a grandcanonical ensemble exchanging energy and particle with the environment.

Proceeding as we did in the case of canonical ensembles we obtain for the probability of a N-particle system in the n-th state

$$
\begin{aligned}
w_{nN} &= Constant \times \exp\left(\frac{S\left(E_0 - E_{nN}, N_0 - N\right)}{k_B}\right) \\
&= A\exp\left(\frac{\mu N - E_{nN}}{k_B T}\right)
\end{aligned}
\tag{4.4.19}
$$

The normalization of probability $\sum_{nN} w_{nN} = 1$ allows us to calculate

$$
\frac{1}{A} = \sum_N \left[\exp\left(\frac{\mu N}{k_B T}\right)\right] \sum_n \left[\exp\left(-\frac{E_{nN}}{k_B T}\right)\right].
\tag{4.4.20}
$$

Using the corresponding quantum mechanical operators we write the grandcanonical density matrix

$$
\hat{\rho}_{\mathrm{gc}} = \frac{\exp\left(-\frac{\hat{\mathcal{H}} - \mu\hat{N}}{k_B T}\right)}{Tr\left(\exp\left(-\frac{\hat{\mathcal{H}} - \mu\hat{N}}{k_B T}\right)\right)}.
\tag{4.4.21}
$$

The operator $\hat{\mathcal{H}} - \mu\hat{N}$ appears so often in condensed matter physics that people working there sometimes use the term **thermodynamic hamiltonian**

$$
\hat{\mathcal{H}}_{\mathrm{therm}} = \hat{\mathcal{H}} - \mu\hat{N}.
\tag{4.4.22}
$$

The **grandcanonical partition function** Z_{gc} is now defined as

$$
Z_{\mathrm{gc}}(T, V, \mu) \overset{\mathrm{def}}{=} \sum_{n,N} \exp\left(\frac{1}{k_B T}\left(\mu N - E_{n,N}\right)\right)
\tag{4.4.23}
$$

$$
= \sum_{N=0}^{\infty} \lambda^N Z_N(T, V)
\tag{4.4.24}
$$

$$
= \frac{1}{A}
\tag{4.4.25}
$$

where

$$
\lambda = \exp\left(\frac{\mu}{k_B T}\right)
\tag{4.4.26}
$$

is called the **fugacity** of the system, and

$$
Z_N(T, V) = \sum_n \exp\left(-\frac{E_{n,N}}{k_B T}\right)
\tag{4.4.27}
$$

is the partition function of a canonical ensemble with fixed number of particle N.

The definition of entropy, Equation 3.4.9 from Boltzmann's principle

$$S = -k_B \sum_{n,N} w_{n,N} \log w_{n,N} \tag{4.4.28}$$

will allow us to obtain

$$-k_B T \log Z_{\text{gc}} = k_B T \log A = \overline{E} - TS - \mu \overline{N} = F - \mu \overline{N}. \tag{4.4.29}$$

We get a new thermodynamic potential

$$\Omega = F - \mu N = -k_B T \log Z_{\text{gc}}(T, V, \mu), \tag{4.4.30}$$

and using Equation 4.4.1 we obtain the *Equation of State* for a grandcanonical ensemble

$$PV = k_B T \log Z_{\text{gc}}(T, V, \mu). \tag{4.4.31}$$

The functional dependence of Ω and the average number of particle $\overline{N}$ are obtained from

$$\begin{aligned} d\Omega &= dF - d(\mu \overline{N}) \\ &= -S dT - P dV - \overline{N} d\mu. \end{aligned} \tag{4.4.32}$$

It is clear that

$$\Omega \equiv \Omega(T, V, \mu), \tag{4.4.33}$$

$$\overline{N} = -\left(\frac{\partial \Omega}{\partial \mu}\right)_{T,V}. \tag{4.4.34}$$

In terms of Ω we can write the **grandcanonical distribution function** as

$$w_{n,N} = \exp\left(\frac{\Omega + \mu N - E_{n,N}}{k_B T}\right). \tag{4.4.35}$$

4.4.3 *Density Fluctuation in Grandcanonical Ensemble and Equivalence of Grandcanonical and Canonical Ensembles*

We calculated the fluctuation of Energy for a canonical ensemble and showed that for macroscopic bodies the relative fluctuation is infinitesimally small and thus energy can be treated as constant. In the same spirit we now calculate the fluctuation in number of particle N for a grandcanonical ensemble where the subsystems exchange particles among themselves.

The average number of particle of a subsystem is

$$\overline{N} = \sum_{n,N} N w_{n,N} = k_B T \frac{\partial}{\partial \mu} \log Z_{gc} = k_B T \frac{Z_{gc}{}'}{Z_{gc}}, \qquad (4.4.36)$$

where we have used the notation $(\cdots)' = \frac{\partial}{\partial \mu}(\cdots)$. We also have

$$\overline{N^2} = \sum_{n,N} N^2 w_{n,N} = (k_B T)^2 \frac{Z_{gc}{}''}{Z_{gc}}, \qquad (4.4.37)$$

so that we obtain for the mean square fluctuation of the number of particle

$$\overline{(\Delta N)^2} = \overline{N^2} - \overline{N}^2 = (k_B T)^2 \frac{\partial}{\partial \mu} \left(\frac{Z_{gc}{}'}{Z_{gc}} \right) = k_B T \left(\frac{\partial \overline{N}}{\partial \mu} \right)_{T,V} \propto \overline{N}. \quad (4.4.38)$$

Hence the relative fluctuation of the number of particle is $(\Delta N)_{rel} \propto \frac{1}{\sqrt{\overline{N}}}$, and for macroscopic bodies the number of particles N of a grandcanonical ensembles may be taken as constant like in a canonical ensemble. In this sense the grandcanonical esemble is said to be equivalent to the canonical ensemble.

4.5 Problems

Problem 4.1.   Construct the distribution function for an ensemble whose subsystems exchange heat at constant pressure (instead of energy) among themselves.

Problem 4.2.   Obtain expressions for the Partition Function and Helmholtz' Free Energy for a system of N spin-$\frac{1}{2}$ particles with intrinsic magnetic moment m in an external magnetic field B.

Problem 4.3.   Considering the magnetic moment m in Problem 4.2 as the external parameter obtain expression for the generalized force.

Problem 4.4.   Obtain the equation of state for a 2-dimensional thin film.

Problem 4.5.   Obtain the equation of state for a 1-dimensional linear polymer.

Chapter 5

Ideal Gas

5.1 Boltzmann Distribution

An **Ideal Gas**, also called a **Perfect Gas** is by definition a system where the interparticle interaction can be neglected. This situation is true when the number density $\frac{N}{V}$ in the system is very small. In this case almost all the single particle states are empty, as there are few particles around to occupy them. The resulting **Boltzmann distribution** is then characterized by a very small value for the *average occupancy* in each level

$$\overline{n_k} \ll 1, \quad \text{for all } k. \tag{5.1.1}$$

If the total number of gas molecule is N, we can treat each molecule to be a subsystem and the total energy is given by

$$E_{\{n_k\}} = \sum_k n_k \epsilon_k, \tag{5.1.2}$$

$$N = \sum_k n_k. \tag{5.1.3}$$

Total energy depends on the number distribution $\{n_1, n_2, \cdots, n_k, \cdots\}$ in the energy levels $\{\epsilon_1, \epsilon_2, \cdots, \epsilon_k, \cdots\}$.

Using Equations 4.4.23 & 4.4.30 we obtain

$$
\exp\left(-\frac{\Omega}{k_B T}\right) = \sum_{N=0}^{\infty} \sum_{n_1+n_2+\cdots=N} \exp\left(\frac{1}{k_B T}[n_1(\mu-\epsilon_1)+n_2(\mu-\epsilon_2)+\cdots]\right)
$$

$$
= \sum_{n_1=0}^{\infty} \sum_{n_2=0}^{\infty} \cdots \exp\left(\frac{1}{k_B T}[n_1(\mu-\epsilon_1)+n_2(\mu-\epsilon_2)+\cdots]\right)
$$

$$
= \prod_k \sum_{n_k=0}^{\infty} \exp\left(\frac{1}{k_B T}[n_k(\mu-\epsilon_k)]\right). \tag{5.1.4}
$$

In arriving at the above result we have used Equation 14.5.1 of § 14.5. We can then write

$$\Omega = \sum_k \Omega_k, \tag{5.1.5}$$

where

$$\Omega_k = -k_B T \log \sum_{n_k=0}^{\infty} \exp\left(\frac{1}{k_B T}\left[n_k \left(\mu - \epsilon_k\right)\right]\right), \tag{5.1.6}$$

and the average number of particle $\overline{n_k}$ in the kth single particle state is

$$\overline{n_k} = -\left(\frac{\partial \Omega_k}{\partial \mu}\right)_{T,V}. \tag{5.1.7}$$

The density matrix takes the form

$$w_{\{n_k\}} = \prod_k w_{n_k} \tag{5.1.8}$$

with

$$w_{n_k} = \exp\left(\frac{1}{k_B T}\left[n_k \left(\mu - \epsilon_k\right)\right]\right). \tag{5.1.9}$$

Since Boltzmann distribution is defined as one with $\overline{n_k} \ll 1$ for all k, most of the levels are unoccupied and

$$w_{n_k=0} = \exp\left(\frac{\Omega_k}{k_B T}\right) \approx 1, \tag{5.1.10}$$

and

$$\overline{n_k} = \sum_{n_k=0}^{\infty} n_k w_{n_k} \approx 1 \cdot w_{n_k=1} = \exp\left(\frac{\mu - \epsilon_k}{k_B T}\right). \tag{5.1.11}$$

5.2   Partition Function, Free Energy and Equation of State

In order to obtain the expression for the partition function Z_N for a system with fixed number of particle $N = \sum_k n_k$ and energy $E = \sum_k n_k \epsilon_k$

$$Z_N = \sum_{\substack{\text{distinct states} \\ n_1 + n_2 \cdots = N}} \prod_k \exp\left(-\frac{n_k \epsilon_k}{k_B T}\right), \tag{5.2.1}$$

we remind us that each particle constitutes an independent subsystem, and hence for the total system of an ideal gas we can write

$$Z_{\text{id}} = \frac{1}{N!} \left[\sum_k \exp\left(-\frac{\epsilon_k}{k_B T} \right) \right]^N, \tag{5.2.2}$$

where the prescription for avoiding Gibbs' paradox has been followed. The Helmholtz' Free Energy becomes

$$F_{\text{id}} = -k_B T \log Z_{\text{id}} = -N k_B T \log \left[\frac{e}{N} \sum_k \exp\left(-\frac{\epsilon_k}{k_B T} \right) \right], \tag{5.2.3}$$

where e is the base of Naperian logarithm.

The energy ϵ_k for each constituent particle of mass m and momentum p has two components, one due to the translational motion $\epsilon_{\text{trans}} = \frac{p^2}{2m}$ and the other due to internal motion ϵ_n^{int}. This internal energy may be due to vibrational or rotational motion if the constituents are polyatomic molecules, or due to internal quantum excitations of the constituent atoms. The Free energy thus has two parts

$$F_{\text{id}} = F_{\text{trans}} + F_{\text{int}}, \tag{5.2.4}$$

where

$$F_{\text{trans}} = -N k_B T \log \left[\frac{e g_J V}{N (2\pi\hbar)^3} \int e^{-\frac{p^2}{2m k_B T}} 4\pi p^2 dp \right]$$

$$= -N k_B T \log \left[\frac{e g_J V}{N} \left(\frac{m k_B T}{2\pi\hbar^2} \right)^{3/2} \right], \tag{5.2.5}$$

$$F_{\text{int}} = -N k_B T \log \sum_n \exp\left(-\frac{\epsilon_n^{\text{int}}}{k_B T} \right). \tag{5.2.6}$$

Here g_J is the degeneracy of the atomic constituents due to their intrinsic angular momentum J. If we separate the volume dependence in the expression for free energy we can write

$$F_{\text{id}} = -N k_B T \log \left(\frac{eV}{N} \right) + N f(T), \tag{5.2.7}$$

$$f(T) = -k_B T \log \left[g_J \left(\frac{m k_B T}{2\pi\hbar^2} \right)^{3/2} \right] - k_B T \log \sum_n \exp\left(-\frac{\epsilon_n^{\text{int}}}{k_B T} \right). \tag{5.2.8}$$

We immediately obtain for pressure

$$P_{id} = -\left(\frac{\partial F_{id}}{\partial V}\right)_{T,N} = \frac{Nk_BT}{V} \qquad (5.2.9)$$

which leads to the equation of state known as the **Perfect Gas Law**.

$$P_{id}V = Nk_BT. \qquad (5.2.10)$$

The other thermodynamic functions follow easily:

$$\text{Gibbs' Potential}: G_{id} = F_{id} + P_{id}V$$
$$= -Nk_BT\log\left(\frac{eV}{N}\right) + Nf(T) + Nk_BT$$
$$= Nk_BT\log P_{id} + N\chi(T), \qquad (5.2.11)$$

where

$$\chi(T) = f(T) - k_BT\log(k_BT), \qquad (5.2.12)$$

$$\text{Entropy}: S_{id} = -\left(\frac{\partial F_{id}}{\partial T}\right)_{V,N}$$
$$= Nk_B\log\left(\frac{eV}{N}\right) - Nf'(T), \qquad (5.2.13)$$

$$\text{Internal Energy}: E_{id} = F_{id} + TS_{id}$$
$$= Nf(T) - NTf'(T), \qquad (5.2.14)$$

$$\text{Heat Capacity}: C_v = \left(\frac{\partial E_{id}}{\partial T}\right)_{V,N}$$
$$= -NTf''(T) = Nc_v. \qquad (5.2.15)$$

Here c_v is the specific heat and prime on a function means derivative of this function with respect to the temperature T.

For a wide range of temperature c_v is constant and Equation 5.2.15 can be integrated to give $f(T)$ as

$$f(T) = \epsilon_0 - \zeta k_BT - c_vT\log(k_BT). \qquad (5.2.16)$$

Here ϵ_0 and the **chemical constant** ζ are the two constants of integration.

5.3 Specific Heat: Translational, Vibrational and Rotational Components

Measurement of specific heat and its variation with temperature is perhaps the most important static thermal measurements on a system. These data establish the nature of internal dynamics of a system. We now investigate the expression of specific heat obtained from application of Boltzmann statistics.

(i) *Monatomic system*: In this case there are no internal motion of the system and rigid translation is the only contributor to specific heat. Here

$$f(T) = -k_B T \log \left[g_J \left(\frac{mk_B T}{2\pi \hbar^2} \right)^{3/2} \right]$$

$$= -\frac{3}{2} k_B T \log (k_B T) - k_B T \log \left[g_J \left(\frac{m}{2\pi \hbar^2} \right)^{3/2} \right], \quad (5.3.1)$$

and comparing this with Equation 5.2.16 we get

$$Specific\ heat: c_{\text{trans}} = \frac{3}{2} k_B, \quad (5.3.2)$$

$$Chemical\ constant: \quad \zeta = \log \left[g_J \left(\frac{m}{2\pi \hbar^2} \right)^{3/2} \right]. \quad (5.3.3)$$

This expression for specific heat is in accord with the law of **Equipartition of Energy** which states that each degree of freedom of the system contributes $\frac{1}{2} k_B T$ to the internal energy. Thus each degree of freedom contributes $\frac{1}{2} k_B$ to the specific heat.

It should be mentioned that Law of Equipartition of Energy is a consequence of Boltzmann statistics and is not valid for quantum statistics.

(ii) *Polyatomic system*: For polyatomic systems there are two modes of internal motion that contribute to the hamiltonian and thus to specific heat. The atoms constituting the molecule execute vibrations about the centre of inertia of the molecule and the molecule executes rigid rotation about its centre of inertia. For both the cases it is easier to handle the quantum mechanical expression of energy states. It is worth remembering that only the energy levels are considered quantum mechanically, but the statistics is still classical.

(a) Molecular Vibration. For simplicity we consider the system of diatomic molecules having only one degree of freedom for the internal vibrational mode of frequency ω. Quantum mechanical energy for the quantum number v is

$$\epsilon_v^{\text{int}} = \hbar\omega \left(v + \frac{1}{2} \right). \tag{5.3.4}$$

From Equation 5.2.6 we get the expression of Free Energy due to vibration of the system as

$$F_{\text{vib}} = -k_B T \log \left(\left[\sum_{v=0}^{\infty} \exp \left(-\frac{\hbar\omega \left(v + \frac{1}{2} \right)}{k_B T} \right) \right]^N \right)$$

$$= -N k_B T \log \left[\frac{\exp \left(-\frac{\hbar\omega}{2k_B T} \right)}{1 - \exp \left(-\frac{\hbar\omega}{k_B T} \right)} \right]$$

$$= \frac{N\hbar\omega}{2} + N k_B T \log \left[1 - \exp \left(-\frac{\hbar\omega}{k_B T} \right) \right]. \tag{5.3.5}$$

From Equation 5.3.5 we easily obtain expressions for entropy, internal energy and heat capacity due to vibrational motion of the system

$$S_{\text{vib}} = -\left(\frac{\partial F_{\text{vib}}}{\partial T} \right)$$

$$= -N k_B \log \left[1 - \exp \left(-\frac{\hbar\omega}{k_B T} \right) \right] + \frac{N\hbar\omega}{T} \frac{1}{\exp \left(\frac{\hbar\omega}{k_B T} \right) - 1}, \tag{5.3.6}$$

$$E_{\text{vib}} = F_{\text{vib}} + T S_{\text{vib}}$$

$$= \frac{N\hbar\omega}{2} + \frac{N\hbar\omega}{\exp \left(\frac{\hbar\omega}{k_B T} \right) - 1}, \tag{5.3.7}$$

$$C_{\text{vib}} = \left(\frac{\partial E_{\text{vib}}}{\partial T} \right)$$

$$= N k_B \left(\frac{\hbar\omega}{k_B T} \right)^2 \frac{\exp \left(\frac{\hbar\omega}{k_B T} \right)}{\left[\exp \left(\frac{\hbar\omega}{k_B T} \right) - 1 \right]^2}. \tag{5.3.8}$$

In Figure 5.3.1 we have plotted the heat capacity of a system as a function of temperature arising from vibrational mode. Though Equation 5.3.8 is an exact expression valid for all range of temperature, it will be helpful to have the asymptotic form for both the low and the high temperature region.

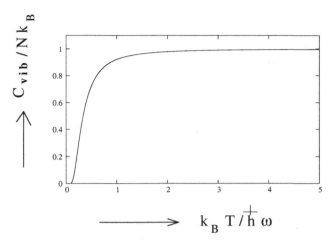

Fig. 5.3.1 Plot of the vibrational component of heat capacity of a 1-dimensional harmonic oscillator of frequency ω as a function of temperature, all in characteristic units of the system.

In the *low temperature* region when the thermal energy k_BT is much less than the characteristic energy $\hbar\omega$ of the system, *i.e.* $k_BT \ll \hbar\omega$ heat capacity has the form

$$C_{\text{vib}} = Nk_B \left(\frac{\hbar\omega}{k_BT} \right)^2 \exp\left(-\frac{\hbar\omega}{k_BT} \right). \tag{5.3.9}$$

In the *high temperature* region when $k_BT \gg \hbar\omega$ we have

$$C_{\text{vib}} = Nk_B, \tag{5.3.10}$$

consistent with the law of equipartition of energy.

In general polyatomic molecules will have more than one normal modes of vibration of different frequencies ω_α. An analysis similar to one described above will give the expression of total Heat Capacity due to all the modes of vibrations as

$$C_{\text{vib}} = \sum_\alpha Nk_B \left(\frac{\hbar\omega_\alpha}{k_BT} \right)^2 \frac{\exp\left(\frac{\hbar\omega_\alpha}{k_BT} \right)}{\left[\exp\left(\frac{\hbar\omega_\alpha}{k_BT} \right) - 1 \right]^2}. \tag{5.3.11}$$

(b) **Molecular Rotation.** The other important internal motion is the rigid rotation of a molecule. *Uniaxial molecules* with moment of inertia I in the rotational quantum state K of degeneracy $(2K+1)$ has the energy

$$\epsilon_K^{\text{int}} = \frac{(\text{Angular Momentum})^2}{2 \times (\text{Moment of Inertia})} = \frac{K(K+1)\hbar^2}{2I}, \qquad (5.3.12)$$

and the contribution to Free Energy due to the rigid rotation of the molecule becomes

$$F_{\text{rot}} = -Nk_BT \log \left[\sum_{K=0}^{\infty} (2K+1) \exp \left(-\frac{K(K+1)\hbar^2}{2Ik_BT} \right) \right]. \tag{5.3.13}$$

We now consider the two limiting cases of temperature separately. In the *low temperature region* when the thermal energy k_BT is much less than $\frac{\hbar^2}{I}$, the characteristic energy of the system, *i.e* $k_BT \ll \frac{\hbar^2}{I}$, contributions to the summation in F_{rot} will come only from small values of K. Thus

$$\lim_{T \to 0} F_{\text{rot}} = -Nk_BT \log \left[1 + 3\exp \left(-\frac{\hbar^2}{Ik_BT} \right) \right]$$

$$= -3Nk_BT \exp \left(-\frac{\hbar^2}{Ik_BT} \right), \qquad (5.3.14)$$

and the form of the heat capacity in this temperature range is

$$C_{\text{rot}} = 3Nk_B \left(\frac{\hbar^2}{Ik_BT} \right)^2 \exp \left(-\frac{\hbar^2}{Ik_BT} \right). \tag{5.3.15}$$

In the other limiting case of *high temperature* when $k_BT \gg \frac{\hbar^2}{I}$, only large values of K will contribute to the summation appearing in Equation 5.3.13, so that summation there can be replaced by integration

$$\sum_{K=0}^{\infty} (2K+1) \exp \left(-\frac{K(K+1)\hbar^2}{2Ik_BT} \right) = \int_0^{\infty} (2K+1)$$

$$\times \exp \left(-\frac{K(K+1)\hbar^2}{2Ik_BT} \right) dK$$

$$\approx \left(\frac{2Ik_BT}{\hbar^2} \right) \int_0^{\infty} e^{-t} dt,$$

$$\tag{5.3.16}$$

Free Energy due to rotational motion takes the form

$$F_{\rm rot} = -Nk_BT \log\left(\frac{2Ik_BT}{\hbar^2}\right). \qquad (5.3.17)$$

and the high temperature heat capacity due to rotational motion becomes

$$C_{\rm rot} = Nk_B, \qquad (5.3.18)$$

again consistent with the law of Equipartition of energy.
In Figure 5.3.2 we have plotted a schematic representation of the Heat capacity $C_{\rm rot}$ due to rotational motion as a function of temperature T.

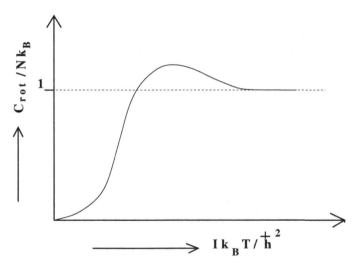

Fig. 5.3.2 Schematic plot of the rotational component of heat capacity of a system of uniaxial molecules with Moment of Inertia I as a function of temperature, all in characteristic units of the system.

5.4 Degeneracy Temperature

Whenever an approximation is used in a theory of physics, we must have an idea when this approximation makes sense. We now make an estimate of the temperature domain where the approximation of Boltzmann statistics $\overline{n_k} \ll 1$, for all k is a valid description of a physical system. Since this is an estimate, we can consider a monatomic system with only translational degree of freedom and ignore all complications like internal excitations. Using Equation 5.1.11 for $\epsilon_k = 0$ we get the condition of applicability of Boltzmann distribution as

$$\exp\left(\frac{\mu}{k_B T}\right) \ll 1. \tag{5.4.1}$$

We immediately conclude that in this case the chemical potential $\mu < 0$. From Equation 5.2.11 for the Gibbs' potential we have

$$\frac{\mu}{k_B T} = \log\left[\frac{N}{g_J V}\left(\frac{2\pi\hbar^2}{m k_B T}\right)^{3/2}\right]. \tag{5.4.2}$$

The restriction on temperature T for the applicability of Boltzmann distribution thus becomes

$$T \gg T_{\deg}, \quad \textbf{the degeneracy temperature} \tag{5.4.3}$$

where

$$T_{\deg} = \frac{2\pi\hbar^2}{m k_B}\left(\frac{N}{g_J V}\right)^{2/3}. \tag{5.4.4}$$

A very good example of non-interacting system is the *free electrons* in the noble metals Cu, Ag, and Au. The number density $\frac{N}{V}$ of electrons in a chemical of Molecular weight M, valency v and density ρ can be evaluated as follows. M grams of the system has the Avogadro Number N_A of molecules each contributing v number of electrons contained in a volume $\frac{M}{\rho}$. Thus

$$\frac{N}{V} = \frac{N_A \rho v}{M}. \tag{5.4.5}$$

For metallic Cu

Density:	ρ	$= 8.93 \ g/cm^3$,
Molecular Weight:	M	$= 63.54$,
Valency:	v	$= 1$,
Spin degeneracy:	g_J	$= 2$.

Using the values of the fundamental constants

Avogadro Constant:	N_A	$= 6.022 \times 10^{23}/mol$,
Planck's Constant:	$\hbar$	$= 1.054 \times 10^{-27} erg - sec$,
Boltzmann Constant:	k_B	$= 1.38 \times 10^{-16} erg/K$,
Electronic Mass:	m_e	$= 9.1 \times 10^{-28} g$,

we obtain the value of the degeneracy temperature

$$T_{\text{deg}} = 6.750 \times 10^4 \ K.$$

This high value of the degeneracy temperature is true for all metallic elements. This clarifies why the values of the Lorentz ratio of thermal conductivity to electrical conductivity as also the value of thermionic emission currents from metals could only be explained on the basis of quantum statistical distribution and *not* from classical Boltzmann distribution.

5.5 Problems

Problem 5.1. Calculate the translational part of Helmholtz' Free Energy F_{trans}, surface tension S, Heat Capacity C_v and Gibbs' Potential G for a system of N free particles, each of mass m, moving in a 2-dimensional thin film of area A.

Problem 5.2. Calculate the translational part of Helmholtz' Free Energy F_{trans}, tension $\mathcal{F}$, Heat Capacity C_v and Gibbs' Potential G for a system of N free particles, each of mass m, moving in a 1-dimensional linear polymer of length L.

Problem 5.3. Obtain expression for degeneracy temperature T_{deg} for a system of N free particles, each of mass m, moving in a 2-dimensional thin film of area A.

Problem 5.4. Obtain expression for degeneracy temperature T_{deg} for a system of N free particles, each of mass m, moving in a 1-dimensional linear polymer of length L.

Chapter 6

Chemical Reaction Equilibrium

6.1 Conditions of Chemical Equilibrum

In the last chapter we have studied systems for which the number of particles remain constant. We now investigate processes in which the numbers of the constituents change. We call them by the generic name of **Chemical Reaction**. They include not only the conventional chemical reactions like $2H_2 + O_2 \rightleftharpoons 2H_2O$ but also fundamental processes like pair annihilation/creation $e^- + e^+ \rightleftharpoons 2\gamma$ or solid state processes like electron-hole creation/destruction $n + p \rightleftharpoons$ photon/phonon quanta. We also point out at the very beginning that the final state of chemical equilibrium is independent of how it is reached, e.g. presence of catalyst *etc*. The processes *during* establishment of equilibrium comes under what is known as **Chemical Kinetics** with which we are not concerned here. We are investigating the state *after* equilibrium has been established.

To understand the processes that lead to the establishment of chemical equilibrium, we take the example of water formation from H_2 and O_2,

$$2H_2 + O_2 \rightleftharpoons 2H_2O. \tag{6.1.1}$$

One starts with a mixture of Hydrogen and Oxygen and initiates the reaction. There are no water molecule at the beginning. As the rate of reaction is related to the concentrations of the constituent molecules and increases with it, the forward reaction (the process from left to right) is predominant to start with and the concentration of water molecule starts growing at the cost of those of H_2 and O_2. After a while backward reaction starts building up. Finally, chemical equilibrium is established, when the concentrations of the constituents taking part in the reaction process become steady.

Generalizing the notations to denote any chemical reaction, we write this as follows

$$\sum_i \nu_i A_i = 0, \tag{6.1.2}$$

where A_i is the $i-th$ chemical species taking part in the reaction and $\nu_i =$ positive or negative integers including 0 is the number of molecules needed to balance the equation. As an example, in the reaction Equation 6.1.1,

$$A_1 = H_2, \ A_2 = O_2, \ A_3 = H_2O, \ \text{and} \tag{6.1.3}$$

$$\nu_1 = +2, \ \nu_2 = +1, \ \nu_3 = -2. \tag{6.1.4}$$

Similarly we can write the *ionization* and the *particle-hole annihilation* reactions

$$A \rightleftharpoons A^+ + e^- \tag{6.1.5}$$

$$n^- + p^+ \rightleftharpoons \gamma \tag{6.1.6}$$

as

$$A - A^+ - e^- = 0, \tag{6.1.7}$$

$$n^- + p^+ - \gamma = 0. \tag{6.1.8}$$

When equilibrium is established at *constant temperature T and pressure P*, the Gibbs' potential G is minimum, so that

$$\sum_i \frac{\partial G}{\partial N_i} \delta N_i = 0, \tag{6.1.9}$$

and hence

$$\sum_i \nu_i \mu_i = 0, \tag{6.1.10}$$

Here $\mu_i = \frac{\partial G}{\partial N_i}$ is the chemical potential and δN_i is the change in number of the $i-th$ species taking part in the reaction. Equation 6.1.10 is the **condition of equilibrium of the chemical reaction**.

In many reactions involving organic compounds as also in some nuclear processes, the reaction proceeds through different partial reactions

$$\sum_i \nu_{i\alpha} A_i = 0, \quad \text{for each } \alpha \text{ separately,} \tag{6.1.11}$$

with the same restrictions on the values of $\nu_{i\alpha}$. Then the conditions of equilibrium of chemical reactions will be

$$\sum_i \nu_{i\alpha} \mu_i = 0, \quad \text{for each } \alpha \text{ separately.} \tag{6.1.12}$$

6.2   Law of Mass Action

We now proceed to calculate the equilibrium concentrations of different constituents. Using the subscript zero to mean the equilibrium value and omitting the appelate 'id' of the Equation 5.2.11, we can write the condition of chemical equilibrium as

$$\sum_i \nu_i \mu_i \equiv k_B T \sum_i \nu_i \log P_{i0} + \sum_i \nu_i \chi_i(T) = 0, \qquad (6.2.1)$$

where P_{i0} is the equilibrium value of the partial pressure of the $i - th$ constituent. We thus obtain

$$\prod_i P_{i0}^{\nu_i} = K_p(T), \quad \text{where} \qquad (6.2.2)$$

$$K_p(T) = \exp\left(-\frac{\sum_i \nu_i \chi_i(T)}{k_B T}\right) \qquad (6.2.3)$$

If we use the *law of partial pressure*

$$P_{i0} = P c_{i0}, \qquad (6.2.4)$$

where c_{i0} is the equilibrium concentration of the $i - th$ constituent, we can transform the Equations 6.2.2 and 6.2.3 in terms of the equilibrium concentrations

$$\prod_i c_{i0}^{\nu_i} = K_c(P, T), \quad \text{where} \qquad (6.2.5)$$

$$K_c(P, T) = K_p(T) P^{-\sum_i \nu_i}. \qquad (6.2.6)$$

Both $K_p(T)$ and $K_c(P, T)$ are called **Chemical Equilibrium Constants** and Equations 6.2.2 and 6.2.5 which are relations between equilibrium partial pressures or concentrations to the above defined chemical equilibrium constants are called **Laws of Mass Action**. This law has often been used in different Condensed Matter physics problems. One famous example is for calculating the concentrations of electrons and holes in both intrinsic and extrinsic semiconductors. In § 6.4 we shall use it to obtain the concentrations of neutral and ionized atoms as also of electrons in an atomic ionization process.

For reactions between *solutes in weak solutions*, Equation 5.2.11 is no longer valid; however we have a similar relation

$$\mu_i = k_B T \log c_i + \psi_i(P, T), \quad \text{where} \tag{6.2.7}$$

$$c_i = \frac{\text{Number of solute particles}}{\text{Number of solvent particles}}, \tag{6.2.8}$$

and thus we have here a relation

$$\prod_i c_{0i}^{\nu_i} = K_c(P.T), \quad \text{where} \tag{6.2.9}$$

$$K_c(P, T) = \exp\left(-\frac{\sum_i \nu_i \psi_i(P, T)}{k_B T}\right), \tag{6.2.10}$$

similar to the law of mass action, but the pressure dependence is now indeterminate.

6.3   Heat of Reaction and Direction of Reaction

All chemical reactions are accompanied by production or absorbtion of heat and for reactions among gaseous constituents there are sometimes change of volume. All these as also the direction of reaction can be inferred from analyzing Equation 6.1.9.

When equilibrium has not been reached, we have

$$\delta G = \sum_i \frac{\partial G}{\partial N_i} \delta N_i = \sum_i \mu_i \delta N_i \tag{6.3.1}$$

After δn number of elementary reactions has occurred we have $\delta N_i = -\nu_i \delta n$ and the change in Gibbs' potential becomes

$$\delta G = -\delta n \sum_i \nu_i \mu_i \tag{6.3.2}$$

$$= -\delta n \left[k_B T \sum_i \nu_i \log P_i + \sum_i \nu_i \chi_i(T) \right] \tag{6.3.3}$$

$$= k_B T \delta n \left[-\sum_i \nu_i \log P_i + \log K_p(T) \right] \tag{6.3.4}$$

$$= k_B T \delta n \left[-\sum_i \nu_i \log c_i + \log K_c(P, T) \right]. \tag{6.3.5}$$

Equation 6.3.5 is applicable also for solutes in weak solutions if the appropriate Chemical Reaction Constant given in Equation 6.2.10 is used.

The reaction proceeds in the forward (*resp.* backward) direction if $\delta G < 0$ (*resp.* $\delta G > 0$). Using Equations 6.3.4 & 6.3.5 we obtain

$$\text{Forward Reaction}: \begin{cases} \log K_p(T) & < \sum_i \nu_i \log P_i, \quad \text{or} \\ \log K_c(P,T) & < \sum_i \nu_i \log c_i \end{cases} . \quad (6.3.6)$$

$$\text{Backward Reaction}: \begin{cases} \log K_p(T) & > \sum_i \nu_i \log P_i, \quad \text{or} \\ \log K_c(P,T) & > \sum_i \nu_i \log c_i \end{cases} . \quad (6.3.7)$$

The values of P_i or c_i are *not* the equilibrium values but those occurring at any moment of time and the direction of reaction also corresponds to that moment of time.

In order to obtain the expression for δQ_P, the **Heat of Reaction at constant pressure**, which is the most common physical condition under which conventional chemical reactions occur, we use the thermodynamic relation

$$\begin{aligned} H = E + PV &= G + TS \\ &= G - T \left(\frac{\partial G}{\partial T} \right)_P \\ &= -T^2 \left[\frac{\partial}{\partial T} \left(\frac{G}{T} \right) \right]_P, \end{aligned} \quad (6.3.8)$$

and Equations 6.3.4 and 6.3.5 and then we finally obtain

$$\begin{aligned} \delta Q_P &= (\delta E + P \delta V)_{P=\text{const.}} \\ &= \delta H \\ &= -T^2 \left[\frac{\partial}{\partial T} \left(\frac{\delta G}{T} \right) \right]_P, \\ &= \begin{cases} -k_B T^2 \delta n \left[\frac{\partial}{\partial T} \log K_p(T) \right]_P. \\ -k_B T^2 \delta n \left[\frac{\partial}{\partial T} \log K_c(P.T) \right]_P. \end{cases} \end{aligned} \quad (6.3.9)$$

For *endothermic reactions*, $\delta Q_P > 0$; thus $\left(\frac{\partial}{\partial T} \log K_c \right)_P < 0$ and the reaction is pushed forward.

For *exothermic reactions* on the other hand, $\delta Q_P < 0$; thus $\left(\frac{\partial}{\partial T} \log K_c \right)_P > 0$ and the reaction is pushed backward.

For calculating the change in volume of the gaseous participants in a chemical reaction we note that in case of ideal gases

$$\delta V = \frac{k_B T}{P} \delta N = -\frac{k_B T}{P} \delta n \sum_i \nu_i. \qquad (6.3.10)$$

Thus if $\sum_i \nu_i = 0$ no change of volume occurs as a result of the chemical reaction. Thus production of gaseous Nitric Oxide $N_2 + O_2 \rightleftharpoons 2NO$ does *not* produce any change of volume, while Ammonia is produced by the reaction $N_2 + 3H_2 \rightleftharpoons 2NH_3$ with a change in volume.

6.4   Ionization Equilibrium

One of the basic astrophysical processes is **thermal ionization** of *atoms* (since the thermal energy $k_B T$ is much greater than the binding energy of molecules and hence the molecules dissociate in their constituent atoms) on stellar surface. This ionization process can also be looked upon as a chemical reaction.

At sufficiently high temperature collisions between the atoms start the ionization process. As the ionization builds up more ions recombine with the free electrons generated as a result of the ionization, thus building up the reaction in the backward direction. Finally equilibrium sets in.

At equilibrium, different fractions of the total number of particles are in different stages of ionization like single, double, *etc.* As we have already mentioned, the chemical compounds are completely dissociated at the high temperature $\sim 10^4 \ K$ on stellar surfaces, and it is sufficient to consider only monatomic gases to be present. We use the notations

$$A_0 = \text{Neutral Atom,}$$
$$A_n \ (n = 1, 2, \cdots) = \text{n} - \text{fold ionized Atom,}$$
$$c_0 = \text{concentration of neutral Atom,}$$
$$c_n \ (n = 1, 2. \cdots) = \text{concentration of n} - \text{fold ionized Atom,}$$
$$c = \text{concentration of free electrons.} \qquad (6.4.1)$$

We now write the ionization reactions as

$$A_0 \rightleftharpoons A_1 + e^-, \qquad (6.4.2)$$
$$A_1 \rightleftharpoons A_2 + e^-, \qquad (6.4.3)$$
$$\cdots\cdots\cdots$$
$$A_{n-1} \rightleftharpoons A_n + e^-. \qquad (6.4.4)$$

Charge neutrality will give us

$$c = 1 \cdot c_1 + 2 \cdot c_2 + \cdots. \qquad (6.4.5)$$

We now use Equation 6.2.5 to write

$$\frac{c_{n-1}}{c_n c} = P K_p^{(n)}(T). \qquad (6.4.6)$$

Here $K_p^{(n)}(T)$ is the chemical equilibrium constant for the process involving ionization of A_{n-1}.

We now proceed to calculate

$$\log K_p^{(n)}(T) = -\frac{1}{k_B T} \left[\chi_{n-1}(T) - \chi_n(T) - \chi_{\text{el}} \right], \qquad (6.4.7)$$

using the generic Equations 5.2.8 and 5.2.12

$$\chi_a(T) = -\frac{5}{2} k_B T \log \left(k_B T \right) - \frac{3}{2} k_B T \log \left(\frac{m_a g_a^{2/3}}{2\pi\hbar^2} \right)$$
$$- k_B T \log \sum_{\text{int}} \exp \left(-\frac{\epsilon_a^{\text{int}}}{k_B T} \right). \qquad (6.4.8)$$

In evaluating the contribution to $\chi_a(T)$ from the internal states, only the ground state ϵ_a is taken into account and since they are $\sim eV$ while at the stellar surfaces $k_B T \sim KeV$ only the first term in the expansion of logarithm is kept. We also must remember that electrons being point particles do *not* have any internal energy levels. Under these conditions we obtain

$$\log K_p^{(n)}(T) = \left[\frac{5}{2} \log \left(k_B T \right) + \frac{3}{2} \log \left(\frac{g_{n-1}^{2/3} M_{n-1}}{2\pi\hbar^2} \right) - \frac{\epsilon_{n-1}}{k_B T} \right]$$
$$- \left[\frac{5}{2} \log \left(k_B T \right) + \frac{3}{2} \log \left(\frac{g_n^{2/3} M_n}{2\pi\hbar^2} \right) - \frac{\epsilon_n}{k_B T} \right]$$
$$- \left[\frac{5}{2} \log \left(k_B T \right) + \frac{3}{2} \log \left(\frac{g_{\text{el}}^{2/3} m_e}{2\pi\hbar^2} \right) \right]$$
$$= -\frac{5}{2} \log \left(k_B T \right) + \frac{3}{2} \log \left[\frac{2\pi\hbar^2 M_{n-1}}{m_e M_n} \left(\frac{g_{n-1}}{g_{\text{el}} g_n} \right)^{2/3} \right]$$
$$+ \frac{\epsilon_n - \epsilon_{n-1}}{k_B T}. \qquad (6.4.9)$$

Here M_k is the mass of the species A_k and ϵ_k is its ground state internal energy for $k = n, n-1$.

Identifying the *Ionization Energy* I_n of the $n - th$ electron from A_{n-1} as

$$I_n = \epsilon_n - \epsilon_{n-1} \qquad (6.4.10)$$

we ultimately obtain

$$K_p^{(n)}(T) = \frac{g_{n-1}}{g_n g_{\text{el}}} \left(\frac{2\pi\hbar^2 M_{n-1}}{m_e M_n} \right)^{3/2} \frac{1}{(k_B T)^{5/2}} \exp\left(\frac{I_n}{k_B T} \right). \qquad (6.4.11)$$

6.5   Saha Formula

The Indian physicist Saha used Equation 6.4.11 to obtain values of temperature of stellar surfaces. Since the binding energy I_1 of the first electron is much less than those of the subsequent electrons, we can consider equilibrium of only neutral atoms, singly ionized atoms and free electrons. Equation 6.4.6 will thus reduce to

$$\frac{c_0}{c_1 c} = P K_p^{(1)}(T), \qquad \text{with} \qquad (6.5.1)$$

$$c = c_1 \qquad \text{because of charge neutrality.} \qquad (6.5.2)$$

In terms of **Degree of Ionization**

$$\alpha = \frac{\text{Number of ionized atoms}}{\text{Total number of atoms and ions}} \qquad (6.5.3)$$

the different concentrations can be written as

$$c = c_1 = \frac{\alpha}{1+\alpha}, \quad \text{and} \quad c_0 = \frac{1-\alpha}{1+\alpha}, \qquad (6.5.4)$$

and Equation 6.5.1 will become

$$\frac{1-\alpha^2}{\alpha^2} = P K_p^{(1)}, \qquad \text{and thus} \qquad (6.5.5)$$

$$\alpha = \frac{1}{\sqrt{1 + P K_p^{(1)}(T)}}. \qquad (6.5.6)$$

Equation 6.5.6 is known as the **Saha Ionization Formula**.

For the ionization reaction

$$A \rightleftharpoons A^+ + e^- \qquad (6.5.7)$$

of an atomic species A of mass M_A with ground state spectroscopic term value $^{2S+1}L_J$ so that $g_A = 2J+1$ and its singly ionized form A^+ with mass

M_{A^+} and spectroscopic term value $^{2S'+1}L'_{J'}$ and degeneracy $g_{A^+} = 2J' + 1$ and the first ionization energy I_1, the relevant chemical reaction constant is

$$K_p^{(1)}(T) = \frac{g_A}{g_{A^+} \cdot g_{el}} \left(\frac{2\pi\hbar^2 M_A}{m_e M_{A^+}} \right)^{3/2} \frac{1}{(k_B T)^{5/2}} \exp\left(\frac{I_1}{k_B T} \right), \qquad (6.5.8)$$

where m_e = electronic mass, and $g_{el} = 2$.

All the factors appearing in $K_p^{(1)}(T)$ can be measured in laboratory. The ratio of intensities of spectral lines of the atomic species A and its ion A^+ in the stellar spectra would give us α_A for A. So the temperature T on stellar surface would have been known if P, the pressure on the stellar surface were known, which unfortunately it is *not*. This difficulty is surmounted by measuring α_A and α_B for two species A and B and P is eliminated from this pair of data giving the temperature T on the stellar surface.

6.6 Problems

Problem 6.1. Obtain the form of the Law of Mass Action, when the chemical reaction proceeds via more than one partial reactions.

Problem 6.2. Prove Equation 6.4.8.

Problem 6.3. Prove Equation 6.4.9.

Problem 6.4. Obtain the Mass Action equation for the volume densities n of electrons of effective mass m_e^* in the conduction band, and p of holes of effective mass m_h^* in the valence band in case of a 3-dimensional non-degenerate intrinsic semiconductor with energy gap E_g.

Problem 6.5. Obtain the Mass Action equation for the surface densities n of electrons of effective mass m_e^* in the conduction band, and p of holes of effective mass m_h^* in the valence band in case of a 2-dimensional non-degenerate intrinsic semiconductor with energy gap E_g.

Chapter 7

Real Gas

7.1 Free Energy, Virial Equation of State

In the case of an ideal gas, the particle density is very small so that the interaction between the molecules can be neglected. By the term **Real Gas**, we mean a system where the interaction between the molecules is taken into account. Depending on the value of the density, single, double or multiple collision terms become important. One of the earliest attempt to incorporate intermolecular interaction was in the van-der-Waal equation of state for real gases. After this initial analytical description of real gases, various other analytical equations of state were advanced by various scientists. Those attempts are no longer continued. Instead, the **Virial Equation of State**

$$PV = Nk_BT \left[1 + B(T) \left(\frac{N}{V} \right) + C(T) \left(\frac{N}{V} \right)^2 + \cdots \right], \qquad (7.1.1)$$

where $B(T)$, $C(T)$, $\cdots$, called the second, third and other **Virial Coefficients** describe the behaviour of the system. The first Virial Coefficient, obviously, has the value 1. The Virial Coefficients express the effect of inter-particle interaction. In Statistical Physics we shall evaluate the Virial Coefficients from the knowledge of the inter-particle interactions. We can also conclude when such a Virial Equation as a power series in the particle density $\frac{N}{V}$ properly describes an interacting system.

It is easier to arrive at the expressions of the Virial Coefficients from the classical expression of energy for a system of N identical interacting molecules each of mass m,

$$E(q, p) = \sum_{i=1}^{N} \frac{p_i^2}{2m} + U(\mathbf{r}_1, \mathbf{r}_2, \cdots, \mathbf{r}_N), \qquad (7.1.2)$$

where $U(\mathbf{r}_1, \mathbf{r}_2, \cdots, \mathbf{r}_N)$ is the inter-particle potential energy of the system. The Partition Function can be written as

$$Z = \frac{1}{N! \, (2\pi\hbar)^{3N}} \int \exp\left(-\frac{E(q,p)}{k_B T}\right) d\Gamma$$

$$= \frac{1}{N!} \frac{1}{(2\pi\hbar)^{3N}} \left[\prod_{i=1}^{N} \int \exp\left(-\frac{p_i^2}{2mk_B T}\right) d^3\mathbf{p}_i\right]$$

$$\times \int \exp\left(-\frac{U(\mathbf{r}_1, \mathbf{r}_2, \cdots, \mathbf{r}_N)}{k_B T}\right) d^3\mathbf{r}_1 d^3\mathbf{r}_2 \cdots d^3\mathbf{r}_N, \quad (7.1.3)$$

after taking proper care to avoid Gibbs' Paradox. Since for ideal gas $U(\mathbf{r}_1, \mathbf{r}_2, \cdots, \mathbf{r}_N) = 0$ and the coordinate space volume integrals would yield V^N, we can write for the Partition Function

$$Z = Z_{\text{id}} \frac{1}{V^N} \int \exp\left(-\frac{U(\mathbf{r}_1, \mathbf{r}_2, \cdots, \mathbf{r}_N)}{k_B T}\right) d^3\mathbf{r}_1 d^3\mathbf{r}_2 \cdots d^3\mathbf{r}_N, \quad (7.1.4)$$

and for the Free Energy

$$F = F_{\text{id}} - k_B T \log \frac{1}{V^N} \int \exp\left(-\frac{U(\mathbf{r}_1, \mathbf{r}_2, \cdots, \mathbf{r}_N)}{k_B T}\right) d^3\mathbf{r}_1 d^3\mathbf{r}_2 \cdots d^3\mathbf{r}_N$$

$$= F_{\text{id}} - k_B T \log \left[1 + \frac{1}{V^N} \int \left(\exp\left(-\frac{U(\mathbf{r}_1, \mathbf{r}_2, \cdots, \mathbf{r}_N)}{k_B T}\right) - 1\right) \right.$$

$$\left. \times \, d^3\mathbf{r}_1 d^3\mathbf{r}_2 \cdots d^3\mathbf{r}_N\right]. \quad (7.1.5)$$

We now assume that the number density $\frac{N}{V}$ is small enough for us to consider *only one binary collision at any fixed moment of time*. The value of the integral on the right hand side of Equation 7.1.5 is thus small compared to 1 and expanding the logarithm we can write

$$F = F_{\text{id}} - k_B T \left[\frac{1}{V^N} \int \left(\exp\left(-\frac{U(\mathbf{r}_1, \mathbf{r}_2, \cdots, \mathbf{r}_N)}{k_B T}\right) - 1\right) \right.$$

$$\left. \times \, d^3\mathbf{r}_1 d^3\mathbf{r}_2 \cdots d^3\mathbf{r}_N\right]. \quad (7.1.6)$$

If the inter-particle potential between the $a-$th and the $b-$th particle

$$v(\mathbf{r}_a, \mathbf{r}_b) = v(\mathbf{r}_a - \mathbf{r}_b) = v(\mathbf{r}), \quad \text{where} \quad (7.1.7)$$

$$\mathbf{r} = \mathbf{r}_a - \mathbf{r}_b, \quad (7.1.8)$$

depends only on their relative coordinate as is the case in almost all situations, we then have

$$U(\mathbf{r}_1, \mathbf{r}_2, \cdots, \mathbf{r}_N) = \sum_{a \neq b} v(\mathbf{r}_a, \mathbf{r}_b) \qquad (7.1.9)$$

and so

$$\frac{1}{V^N} \int \left(\exp \left(-\frac{U(\mathbf{r}_1, \mathbf{r}_2, \cdots, \mathbf{r}_N)}{k_B T} \right) - 1 \right) d^3 \mathbf{r}_1 d^3 \mathbf{r}_2 \cdots d^3 \mathbf{r}_N$$

$$= \frac{N(N-1)}{2} \frac{V^{N-2}}{V^N} \int \left(\exp \left(-\frac{v(\mathbf{r}_1, \mathbf{r}_2)}{k_B T} \right) - 1 \right) d^3 \mathbf{r}_1 d^3 \mathbf{r}_2$$

$$\approx \frac{N^2}{2V^2} \int \left(\exp \left(-\frac{v(\mathbf{r}_1, \mathbf{r}_2)}{k_B T} \right) - 1 \right) d^3 \mathbf{r}_1 d^3 \mathbf{r}_2$$

$$\approx -\frac{N^2}{2V^2} V \int \left(1 - \exp \left(-\frac{v(\mathbf{r})}{k_B T} \right) \right) d^3 \mathbf{r} \ , \quad (7.1.10)$$

since $\frac{N(N-1)}{2} \approx \frac{N^2}{2}$ independent pairs can be chosen out of N particles. The *Free Energy* of a real gas has now the form

$$F = F_{\text{id}} + F_{\text{int}}, \quad \text{with} \qquad (7.1.11)$$

$$F_{\text{int}} = -N k_B T \frac{N}{V} B(T), \quad \text{where} \qquad (7.1.12)$$

$$B(T) = \frac{1}{2} \int \left(1 - \exp \left(-\frac{v(\mathbf{r})}{k_B T} \right) \right) d^3 \mathbf{r}. \qquad (7.1.13)$$

We now obtain for the *Pressure* of a real gas

$$P = -\left(\frac{\partial F}{\partial V} \right)_T$$

$$= -\left(\frac{\partial F_{\text{id}}}{\partial V} \right)_T - \left(\frac{\partial F_{\text{int}}}{\partial V} \right)_T$$

$$= \frac{N k_B T}{V} \left[1 + B(T) \frac{N}{V} \right]. \qquad (7.1.14)$$

Since we have considered only single binary collisions we have obtained the **Virial Equation of State** correct up to the **Second Virial Coefficient** $B(T)$ given by Equation 7.1.13. For larger values of the particle density $\frac{N}{V}$ we have to include contributions from multiple collisions and higher order Virial coefficients would appear. The Second Virial Coefficient $B(T)$ is thus expressed in term of the interparticle potential, and is no longer an empirical quantity.

7.2   Second Virial Coefficient and Applicability of Virial Equation

We now come to the question *when* an interacting system can be described by the Virial Equation of State with the Second Virial Coeffcient given by Equation 7.1.13. The Virial equation of state will exist if the integral in Equation 7.1.13 exists. The *neccessary condition* for this is, of course, $\lim_{r \to \infty} v(r) = 0$ for a spherically symmetric $v(r)$. But this is *not a sufficient condition*. Expanding the exponential and remembering that $d^3\mathbf{r} = 4\pi r^2 dr$ we find that the spherically symmetric potential $v(r)$ must approach zero faster than $\frac{1}{r^3}$ as $r \to \infty$. This automatically excludes the two *most important* potentials of physics **Coulombic** and **Gravitational**. Thus strong electrolytes and cosmological systems do not satisfy a Virial Equation of State of Taylor series expansion in the particle density $\frac{N}{V}$.

Two most frequently used potentials for describing molecular interaction are the **London's 6-12 potential**. also called **Heitler-London Potential**

$$v(r) = V_{\text{London}}(r) = v_0 \left[\left(\frac{r_m}{r} \right)^{12} - 2 \left(\frac{r_m}{r} \right)^6 \right], \qquad (7.2.1)$$

and the **Morse Potential**

$$v(r) = V_{\text{Morse}}(r) = v_0 \left[e^{-2\left(\frac{r}{r_m} - 1\right)} - 2 e^{-\left(\frac{r}{r_m} - 1\right)} \right]. \qquad (7.2.2)$$

Both these potentials approach 0 sufficiently fast as $r \to \infty$, have a negative minimum value $-v_0$ at $r = r_m$ and a positive repulsive core when $r < 2r_0$, so that r_0 may be looked upon as the rigid molecular radius. For the **Heitler-London** Potential $2r_0 = 2^{-\frac{1}{6}} r_m$ while for the **Morse** Potential $2r_0 = (1 - \log 2) r_m$. For molecules interacting via these potentials, there exists a Virial Equation of State described by Equation 7.1.1. In Figure 7.2.1 we have plotted V_{Morse} in units of v_0 as a function of r/r_m, showing clearly the short-range repulsive core for $r < 2r_0$ and the attractive part of the potential for $r > 2r_0$ with a minimum value $-v_0$ at $r = r_m$.

We indicate below, how we can infer the temperature dependence of the Second Virial Coefficient $B(T)$, which we now write in the form

$$B(T) = 2\pi \int_0^{2r_0} \left[1 - \exp\left(-\frac{v(r)}{k_B T} \right) \right] r^2 dr$$

$$+ 2\pi \int_{2r_0}^{\infty} \left[1 - \exp\left(-\frac{v(r)}{k_B T} \right) \right] r^2 dr. \quad (7.2.3)$$

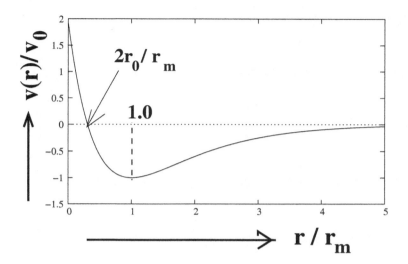

Fig. 7.2.1   Plot of the Morse Potential V_{Morse}/v_0 as a function of r/r_m.

(i) *High Temperature Limit*, $k_B T \gg v_0$: In the region $r < 2r_0$, $v(r)$ is positive and large, so the first integral on the right hand side of Equation 7.2.3 has a *finite positive value*. In the region $r > 2r_0$, $|v(r)| < v_0$ and thus the second integral on the right hand side of Equation 7.2.3 can be written as

$$0 > 2\pi \int_{2r_0}^{\infty} \left[1 - \exp\left(-\frac{v(r)}{k_B T} \right) \right] r^2 dr > -2\pi \int_{2r_0}^{\infty} \frac{|v(r)|}{k_B T} r^2 dr, \quad (7.2.4)$$

and is thus a *very small negative quantity*. So in this limit $B(T) > 0$.

(ii) *Low Temperature Limit*, $k_B T \ll v_0$: In the region $r < 2r_0$ the first integral on the right hand side of Equation 7.2.3 again remains *finite positive*. But the integrand in the second integral on the right hand side of Equation 7.2.3 becomes *a negative large quantity* and so is that integral and thus $B(T) < 0$.

We have schematically drawn in Figure 7.2.2 the dependence of $B(T)$ as a function of temperature T. As we move from low temperature to high temperature $B(T)$ changes sign, and so must pass through the zero value at some temperature T_B, called the **Boyle Temperature**. Since at this temperature the second virial coefficient vanishes the system will behave as an ideal gas even though it is an interacting system.

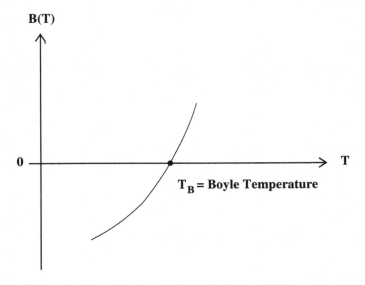

Fig. 7.2.2 Schematic plot of the Second Virial Coefficient $B(T)$ as a function of temperature T.

7.3 Model Calculation and van-der-Waal's Equation of State

A simple model potential which has an infinite hard core of radius $2r_0$ and a short-range attractive constant potential $-v_0$ of finite range

$$v(r) = \begin{cases} \infty & \text{for } r < 2r_0 \\ -v_0 & \text{for } 2r_0 < r < r_1 \\ 0 & \text{for } r > r_1 \end{cases} \qquad (7.3.1)$$

is shown in Figure 7.3.1. We now calculate the second virial coefficient $B(T)$ for this model potential. Using Equation 7.2.3 in the high temperature limit

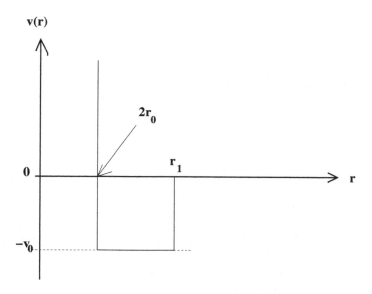

Fig. 7.3.1 A model infinite hard core potential of radius $2r_0$ and a constant attractive part $-v_0$ and range r_1.

$$B(T) = 2\pi \int_0^{2r_0} \left[1 - \exp\left(-\frac{v(r)}{k_B T} \right) \right] r^2 dr$$

$$+ 2\pi \int_{2r_0}^{\infty} \left[1 - \exp\left(-\frac{v(r)}{k_B T} \right) \right] r^2 dr$$

$$= 2\pi \int_0^{2r_0} 1 \cdot r^2 dr + 2\pi \int_{2r_0}^{r_1} \left[1 - \exp\left(\frac{v_0}{k_B T} \right) \right] r^2 dr$$

$$= 2\pi \int_0^{2r_0} 1 \cdot r^2 dr - 2\pi \frac{v_0}{k_B T} \int_{2r_0}^{r_1} 1 \cdot r^2 dr$$

$$= b - \frac{a}{k_B T} \tag{7.3.2}$$

where

$$b = \frac{16\pi r_0^3}{3}, \qquad \text{and} \tag{7.3.3}$$

$$a = \frac{2\pi v_0}{3} \left(r_1^3 - 8r_0^3 \right). \tag{7.3.4}$$

We now use Equation 7.1.14 to obtain

$$P = \frac{Nk_BT}{V}\left[1 + B(T)\frac{N}{V}\right]$$

$$= \frac{Nk_BT}{V} + \frac{Nk_BT}{V}\frac{N}{V}\left(b - \frac{a}{k_BT}\right),$$

$$\left(P + \frac{N^2}{V^2}a\right) = \frac{Nk_BT}{V}\left(1 + \frac{N}{V}b\right)$$

$$\approx \frac{\frac{Nk_BT}{V}}{1 - \frac{N}{V}b},$$

so that we arrive at the **van-der-Waal's Equation of State** for *Real Gases*

$$\left(P + \frac{N^2}{V^2}a\right)(V - Nb) = Nk_BT. \tag{7.3.5}$$

7.4   Joule-Thomson Expansion and Inversion Temperature

In the practical application of liquifaction of gases the physical process of **Joule-Thomson Expansion** has an important role. But in order to utilize this process for liquifying a real gas, the system has to be cooled below the so-called Inversion Temperature T_i. All these facts can be understood from the behaviour of the second virial coefficient.

Starting from the truncated Virial Equation, keeping only the second virial coefficient we can obtain

$$P = \frac{Nk_BT}{V}\left[1 + B(T)\frac{N}{V}\right], \tag{7.4.1}$$

$$\frac{N}{V} = \frac{P}{k_BT}\left[1 + B(T)\frac{N}{V}\right]^{-1}, \tag{7.4.2}$$

$$V = \frac{Nk_BT}{P}\left[1 + B(T)\frac{N}{V}\right] \approx \frac{Nk_BT}{P}\left[1 + B(T)\frac{P}{k_BT}\right]$$

$$= \frac{Nk_BT}{P} + NB(T). \tag{7.4.3}$$

We now describe the essential feature of the **Joule-Thomson Expansion** shown in Figure 7.4.1. Fluid kept at a constant pressure P_1 in the left compartment of volume V_1 is steadily transferred to the right compartment occupying a volume V_2 at a lower constant pressure $P_2 < P_1$ through a

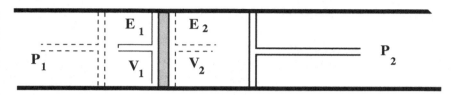

Fig. 7.4.1 Schematic Arrangement of Joule-Thomson Expansion Experiment. The dashed positions of the pistons are those at the beginning of the transfer and the solid positions of the pistons are after the transfer has been complete.

semi-permeable membrane separating the two compartments. The pores of the membrane are so small that the macroscopic velocity of the gas molecules becomes zero, because of the frictional loss at the membrane, making the process an irreversible one. Since no heat has been transferred to the system we have

$$E_2 - E_1 = P_1V_1 - P_2V_2.$$

$$E_1 + P_1V_1 = E_2 + P_2V_2 \tag{7.4.4}$$

so Enthalpy

$$H \equiv E + PV \;\; = \;\; \text{Constant.} \tag{7.4.5}$$

Now

$$dH = TdS + VdP$$
$$= T\left(\frac{\partial S}{\partial T}\right)_P dT + T\left(\frac{\partial S}{\partial P}\right)_T dP + VdP$$
$$= C_P dT - \left[T\left(\frac{\partial V}{\partial T}\right)_P - V\right]dP, \tag{7.4.6}$$

where we have used the definition $C_P = T\left(\frac{\partial S}{\partial T}\right)_P$ of the heat capacity at constant pressure and Maxwell's thermodynamic relation $\left(\frac{\partial S}{\partial P}\right)_T = -\left(\frac{\partial V}{\partial T}\right)_P$.

From $dH = 0$ follows

$$\left(\frac{\partial T}{\partial P}\right)_H = \frac{1}{C_P}\left[T\left(\frac{\partial V}{\partial T}\right)_P - V\right]. \tag{7.4.7}$$

From Equation 7.4.3 we get

$$V = \frac{Nk_BT}{P} + NB(T),$$

$$\left(\frac{\partial V}{\partial T}\right)_P = \frac{Nk_B}{P} + N\frac{dB(T)}{dT},$$

$$T\left(\frac{\partial V}{\partial T}\right)_P - V = N\left[T\frac{dB(T)}{dT} - B(T)\right],$$

$$\left(\frac{\partial T}{\partial P}\right)_H = \frac{N}{C_P}\left[T\frac{dB(T)}{dT} - B(T)\right]. \qquad (7.4.8)$$

For *ideal gases* the right hand side of Equation 7.4.7 is zero and hence Joule-Thomson Expansion does *not* produce any change of temperature.

For *real gas*, however, there is a *change of temperature* as a result of Joule-Thomson expansion. In Figure 7.4.2 we have plotted $T\left(\frac{dB}{dT}\right)_P - B$ as a function of temperature T for a *real gas*. It is seen that the function

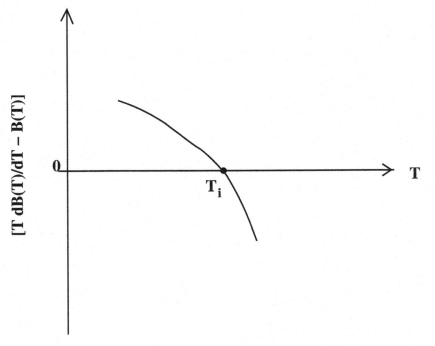

Fig. 7.4.2   Schematic plot of $\left[T\frac{dB(T)}{dT} - B(T)\right]$ as a function of temperature T for a *real gas*.

changes sign at $T = T_i$, called the **Inversion Temperature**. Since in the Joule-Thomson Expansion process $\Delta P < 0$,

(i) For $T < T_i$, *below the Inversion Temperature:* $\Delta T < 0$, *i.e.* the system gets cooled.

(ii) For $T > T_i$, *above the Inversion Temperature:* $\Delta T > 0$, *i.e.* the system gets heated up.

This is the explanation from statistical physics of the fact that a gas has to be pre-cooled below the inversion temperature before it can be liquified by Joule-Thomson expansion.

7.5 Problems

Problem 7.1. Obtain the expression for the 2nd Virial coefficient $B(T)$ and the surface tension S for a 2-dimensional thin film of area A.

Problem 7.2. Obtain the expression for the 2nd Virial coefficient $B(T)$ and the tension $\mathcal{F}$ for a 1-dimensional linear polymer of length L.

Problem 7.3. Show that for a 2-dimensional thin film the inter-particle potential $v(r)$ must go to zero faster than $\frac{1}{r^2}$, as $r \to \infty$, in order that a Virial Equation of state will exist.

Problem 7.4. Show that for a 1-dimensional linear polymer the inter-particle potential $v(r)$ must go to zero faster than $\frac{1}{r}$, as $r \to \infty$, in order that a Virial Equation of state will exist.

Problem 7.5. Obtain the expression for Inversion Temperature T_i for a real gas satisfying van-der-Waal's equation 7.3.5.

Chapter 8

Strong Electrolytes

8.1 Debye-Hückel Approximation, Debye Length

In Chapter 7 we derived the Virial Equation of State for a real gas and also found out that such an analytic expansion in terms of the number density $\frac{N}{V}$ for the pressure P is not valid if the interparticle interaction $v(r)$ tends to zero at large interparticle separation r as slowly as or slower than $\frac{1}{r^3}$. For a strong electrolyte comprising of oppositely charged ions where the interaction is of the Coulombic form $\frac{1}{r}$ Debye and Hückel developed a prescription for deriving the equation of state. For a completely ionized gas or a **plasma** where the particle-particle interaction is also of the Coulombic type the same analysis may be applied.

In an electrolyte, ions of the same charge sign repel each other, while ions of opposite charge signs attract each other creating an **Ion Atmosphere** shown in Figure 8.1.1. In the general case there are many kinds of ions which are denoted by the suffix a. We take $Z_a e$ as the charge of the a−th ion, $e = 4.80325 \times 10^{-10}$ esu being the magnitude of the electronic charge and Z_a with positive or negative non-zero integers being the *ionicity* of the a−th ion. Let n_{a0} be the number of ions of type a per unit volume of the system, then the condition of charge neutrality will require

$$\sum_a Z_a n_{a0} = 0. \tag{8.1.1}$$

Here n_{a0} corresponds to a uniform charge distribution. Debye assumed that the deviation of the system from *idealness* is small, *i.e.* the Kinetic Energy which is of the order of the thermal energy $k_B T$ is much larger than the Coulomb energy $\frac{(Ze)^2}{r}$, where $r \propto n^{-\frac{1}{3}}$ is an average inter-particle separation, Here n is an average particle density and Z is an average ionicity.

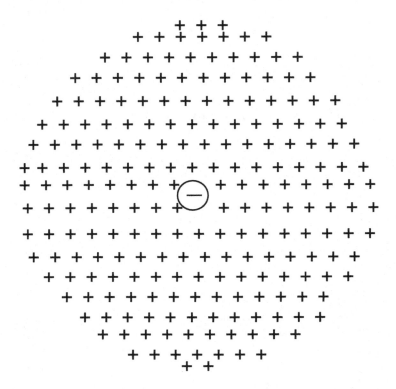

Fig. 8.1.1   Ion Atmosphere comprising of neighbors with opposite charge signs.

Debye's condition thus gives us

$$k_B T \gg \frac{(Ze)^2}{r}, \tag{8.1.2}$$

$$k_B T \gg (Ze)^2 n^{\frac{1}{3}}, \tag{8.1.3}$$

$$n \ll \left(\frac{k_B T}{(Ze)^2}\right)^3, \tag{8.1.4}$$

as the condition on the average density. Because of the smallness of the value of n the system is also sometimes referred to as a **tenuous plasma**.

The energy of the system of uniformly distributed ions assumed to be an ideal gas is taken as the *zero of the energy scale*. The difference of energy between this uniform ion distribution and the actual space dependent ion distribution is called the **correlation energy**

$$E_{\text{corr}} = \frac{V}{2} \sum_a Z_a e n_{a0} \phi_a, \qquad (8.1.5)$$

where ϕ_a is the *electrostatic potential* acting on an ion of type a due to all the other ions. The concept of **ion atmosphere** makes it possible for us to calculate ϕ_a. The ion almosphere is such that each ion creates an inhomogeneously charged ion cloud around itself, which for simplicity's sake we assume to be spherically symmetric. Thus the density distribution of charge around a particular ion a due to all the other ions is assumed to be dependent only on the distance from this ion a. As an example if we choose a particular ion P and n_a is the density of ion of type a around this chosen ion P, then n_a depends on the distance r of ion a from P. Also

$$Z_a e \phi_P = \text{Potential Energy of } a \text{ near P.} \qquad (8.1.6)$$

Hence, using Boltzmann's distribution formula we get

$$n_a = n_{a0} \exp\left(-\frac{Z_a e \phi_P}{k_B T}\right)$$
$$\approx n_{a0} - n_{a0} \frac{Z_a e}{k_B T} \phi_P. \qquad (8.1.7)$$

The constant is put equal to n_{a0}, since at large distance from P $\phi_P \to 0$ and the density of ion of type a becomes the mean ionic density n_{a0}. The electrostatic potential must satisfy Poisson Equation with the total charge density $\sum_a Z_a e n_a$:

$$\nabla^2 \phi_P = -4\pi \sum_a Z_a e n_a$$
$$= \kappa^2 \phi_P, \quad \text{with} \qquad (8.1.8)$$
$$\kappa^2 = \frac{4\pi e^2}{k_B T} \sum_a Z_a^2 n_{a0}, \qquad (8.1.9)$$

where we have made use of Equation 8.1.7 and the charge neutrality Equation 8.1.1. The quantity κ has the physical dimension of inverse of a length and $\frac{1}{\kappa}$ is called **Debye-Hückel Length**.

8.2   Screened Coulomb Potential

For spherically symmetric case Equation 8.1.8 reduces to

$$\frac{1}{r^2}\frac{d}{dr}\left(r^2\frac{d}{dr}\right)\phi_P - \kappa^2\phi_P = 0, \qquad (8.2.1)$$

with solution

$$\phi_P = Z_P e \frac{\exp\left(-\kappa r\right)}{r} \qquad (8.2.2)$$

where the constant of integration is so chosen that very near the ion P with charge $Z_P e$ the potential is coulombic $\frac{Z_P e}{r}$. In Figure 8.2.1 we have plotted $\phi_P(r)$ as a function of r and have compared it with the coulombic function. We see that whereas the $\frac{1}{r}$ potential decreases very slowly, $\phi_P(r)$

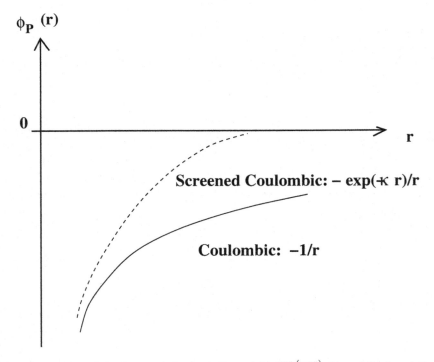

Fig. 8.2.1   Plot of the Screened Coulomb Potential $-\frac{\exp\left(-\kappa r\right)}{r}$ (dotted line) and the Coulomb Potential $-\frac{1}{r}$ (solid line).

decreases very fast to zero. That is the reason that this potential is called the **Screened Coulomb Potential**. This potential is most frequently used for electrons within a solid with a *bare* coulombic interaction among themselves.

We can now make the approximation

$$\phi_P(r) \approx \frac{Z_P e}{r} - Z_P e \kappa. \qquad (8.2.3)$$

The constant term $-Z_P e \kappa$ on the right hand side of Equation 8.2.3 is the extra electrostatic potential produced by the *ion atmosphere*. This *mean field result* justifies the success of **Drude-Sommerfeld free electron theory** in explaining electrical conductivity of metals, even though the electrons interact with long-range coulomb interaction.

8.3 Equation of State and Osmotic Pressure

We are now in a position to arrive at the equation of state for a system with coulombic $\frac{1}{r}$ interaction. In this process we calculate the **correlation energy** E_{corr} of Equation 8.1.5 with ϕ_a replaced by the extra term $-Z_a e \kappa$ obtained in Equation 8.2.3. This is because of our choice of the *zero* of energy. Thus

$$E_{\text{corr}} = \frac{V}{2} \sum_a Z_a e n_{a0} \cdot (-Z_a e \kappa)$$

$$= -V e^3 \sqrt{\frac{\pi}{k_B T}} \left[\sum_a n_{a0} Z_a^2 \right]^{\frac{3}{2}}$$

$$= -e^3 \sqrt{\frac{\pi}{k_B T V}} \left[\sum_a N_a Z_a^2 \right]^{\frac{3}{2}}, \quad \text{where} \qquad (8.3.1)$$

$$N_a = n_{a0} V \qquad (8.3.2)$$

is the totoal number of ions of type a. Using the thermodynamic result

$$\frac{E}{T^2} = \left(\frac{\partial}{\partial T} \left(\frac{F}{T} \right) \right)_V, \qquad (8.3.3)$$

we get the contribution to F, the free energy from the ion atmosphere

$$F_{\text{corr}} = -\frac{2e^3}{3} \sqrt{\frac{\pi}{k_B T V}} \left[\sum_a N_a Z_a^2 \right]^{\frac{3}{2}}, \qquad (8.3.4)$$

so that the total free energy is

$$F = F_{\text{id}} + F_{\text{corr}} \tag{8.3.5}$$

and the total pressure is

$$
\begin{aligned}
P &= -\left(\frac{\partial F}{\partial V}\right)_T \\
&= -\left(\frac{\partial F_{\text{id}}}{\partial V}\right)_T - \left(\frac{\partial F_{\text{corr}}}{\partial V}\right)_T \\
&= \frac{Nk_BT}{V} - \frac{e^3}{3V^{\frac{3}{2}}}\sqrt{\frac{\pi}{k_BT}}\left[\sum_a N_a Z_a^2\right]^{\frac{3}{2}} \tag{8.3.6} \\
&= \frac{Nk_BT}{V} - \frac{e^3}{3}\left(\frac{N}{V}\right)^{\frac{3}{2}}\sqrt{\frac{\pi}{k_BT}}\left[\sum_a c_a Z_a^2\right]^{\frac{3}{2}}, \tag{8.3.7}
\end{aligned}
$$

where c_a is the *concentration* of the ion of type a with $\sum_a c_a = 1$.

We thus arrive at the **Equation of State**

$$PV = Nk_BT\left[1 - \frac{e^3}{3}\sqrt{\frac{\pi}{k_BT}}\left[\sum_a c_a Z_a^2\right]^{\frac{3}{2}}, \left(\frac{N}{V}\right)^{\frac{1}{2}}\right], \tag{8.3.8}$$

for systems like **strong electrolyte** or **tenuous plasma** with coulombic inter-particle interactions, which obviously is not a Taylor Series expansion in terms of $\frac{N}{V}$.

The extra term in the expression Equation 8.3.7 for pressure is the **Osmotic Pressure** in Strong Electrolytes.

8.4   Problems

Problem 8.1.   Show that for a strong electrolyte in a 2-dimensional thin film an isotropic solution for the electrostatic potential with appropriate boundary condition is

$$\phi_P(r) = Ae^{-\kappa r}\sum_{n=0}^{\infty}\frac{\Gamma(2m+1)}{2^m\left(\Gamma(m+1)\right)^3}(\kappa r)^n.$$

Chapter 9

Quantum Statistics

9.1 Bose and Fermi Distributions

We have seen in § 5 that the classical Boltzmann statistics is valid if the average occupancy in any state is much smaller than 1 and the temperature T is higher than the **degeneracy temperature** T_g.

As Equation 5.4.4 shows T_g depends on the number density and we conclude that at *high density and/or at low temperature* quantum statistics and not Boltzmann statistics holds good. There are two quantum statistical distributions, Bose-Einstein(BE) and Fermi-Dirac(FD).

It was first experimentally observed that for particles having even integral multiple of $\frac{\hbar}{2}$ as intrinsic spin, a quantum state can contain any number of particles and the system follows the Bose-Einstein statistics; while for particles having odd integral multiple of $\frac{\hbar}{2}$ as intrinsic spin, the states can contain at most 1 particle and the system follows the Fermi-Dirac statistics. Under certain very general conditions like (i) *causality*, (ii) *local nature of interaction* and (iii) *relativistic invariance* Pauli later concluded that

(i) *from the postulate that the energy must be positive follows the necessity of Fermi-Dirac statistics for particles with arbitrary half-integral spin*, and

(ii) *from the postulate that observables at different space-time points with a space-like distance are commutable follows the necessity of Bose-Einstein statistics for particles with arbitrary integral spin.*

This is really the mathematical conclusions from **Pauli's Exclusion Principle**.

We thus summarize that the applicability of quantum statistics rests on the following criteria:

(i) Temperature T is lower than the degeneracy temperature T_g.

(ii) Integral spin particles obey Bose-Einstein (BE) statistics.

(iii) Odd Half-Integral spin particles obey Fermi-Dirac (FD) statistics.

Particles obeying BE statistics are termed **Bosons**, while those obeying FD statistics are called **Fermions**.

Let us consider a system of N non-interacting particles. The thermo-dynamic potential Ω is given by Equations 5.1.5 and 5.1.6

$$\Omega = \sum_k \Omega_k, \tag{9.1.1}$$

$$\Omega_k = -k_B T \log \sum_{n_k} \left[\exp\left(\frac{n_k (\mu - \epsilon_k)}{k_B T} \right) \right],$$

and the **occupancy** $\overline{n_k}$ in the k-th state is given by Equation 5.1.7

$$\overline{n_k} = -\left(\frac{\partial \Omega_k}{\partial \mu} \right)_T.$$

For **Bose-Einstein statistics** we thus have

$$n_k = 0, 1, 2, \cdots, \tag{9.1.2}$$

$$\Omega_k = +k_B T \log \left[1 - \exp\left(\frac{\mu - \epsilon_k}{k_B T} \right) \right], \tag{9.1.3}$$

$$\overline{n_k} = \frac{1}{\exp\left(\frac{\epsilon_k - \mu}{k_B T} \right) - 1}. \tag{9.1.4}$$

Since the occupancy $\overline{n_k}$ *cannot* be negative even at $T = 0 \ K$ for any value of ϵ_k and in particular for the minimum value $\epsilon_{\min}$, we have the constraint

$$\mu < \epsilon_{\min} \tag{9.1.5}$$

For non-relativistic bosons of mass m, $\epsilon(p) = \frac{p^2}{2m}$ with $\epsilon_{\min} = 0$ and the chemical potential μ will thus satisfy

$$\mu \leq 0. \tag{9.1.6}$$

For relativistic particles with rest mass m, $\epsilon(p) = \sqrt{(pc)^2 + (mc^2)^2}$ has $\epsilon_{\min} = mc^2$ and so for relativistic bosons of rest mass m

$$\mu \leq mc^2. \tag{9.1.7}$$

For **Fermi-Dirac statistics** on the other hand we have

$$n_k = 0, 1, \tag{9.1.8}$$

$$\Omega_k = -k_B T \log \left[1 + \exp \left(\frac{\mu - \epsilon_k}{k_B T} \right) \right], \tag{9.1.9}$$

$$\overline{n_k} = \frac{1}{\exp \left(\frac{\epsilon_k - \mu}{k_B T} \right) + 1}. \tag{9.1.10}$$

If the chemical potential $\mu < \epsilon_{\min}$ then the occupancy $\overline{n_k} = 0$ for all energy levels at temperature $T = 0\ K$, giving rise to a physical contradiction. Thus for fermions

$$\mu > \epsilon_{\min}. \tag{9.1.11}$$

We write for the two statistics in a combined form

$$(\Omega_k)_{\substack{\text{B.E.} \\ \text{F.D.}}} = \pm k_B T \log \left[1 \mp \exp \left(\frac{\mu - \epsilon_k}{k_B T} \right) \right], \tag{9.1.12}$$

$$(\overline{n_k})_{\substack{\text{B.E.} \\ \text{F.D.}}} = \frac{1}{\exp \left(\frac{\epsilon_k - \mu}{k_B T} \right) \mp 1}. \tag{9.1.13}$$

The upper sign on the right hand side corresponds to BE statistics and the lower sign corresponds to the FD statistics. The advantage of writing the combined form is that quite a number of the properties of quantum statistics can be written in a compressed way. The expressions for total number N, total energy E and the thermodynamic potential Ω can thus be written as

$$N_{\substack{\text{B.E.} \\ \text{F.D.}}} = \sum_k (\overline{n_k})_{\substack{\text{B.E.} \\ \text{F.D.}}} = \sum_k \frac{1}{\exp \left(\frac{\epsilon_k - \mu}{k_B T} \right) \mp 1}, \tag{9.1.14}$$

$$E_{\substack{\text{B.E.} \\ \text{F.D.}}} = \sum_k \epsilon_k (\overline{n_k})_{\substack{\text{B.E.} \\ \text{F.D.}}} = \sum_k \frac{\epsilon_k}{\exp \left(\frac{\epsilon_k - \mu}{k_B T} \right) \mp 1}, \tag{9.1.15}$$

$$\Omega_{\substack{\text{B.E.} \\ \text{F.D.}}} = \sum_k (\Omega_k)_{\substack{\text{B.E.} \\ \text{F.D.}}} = \pm \sum_k k_B T \log \left[1 \mp \exp \left(\frac{\mu - \epsilon_k}{k_B T} \right) \right]. \tag{9.1.16}$$

9.2   Quantum Gases of Elementary Particles: Number Density and Chemical Potential, Energy Density, Equation of State

For non-interacting systems the summations over states in Equations 9.1.14, 9.1.15 and 9.1.16 can be replaced by phase space integrals $\frac{4\pi Vg}{(2\pi\hbar)^3}\int_0^\infty (\cdots)\,p^2 dp$, where g is the degeneracy due to the intrinsic spin of the system.

We thus arrive at the expressions for particle density $\frac{N}{V}$, the energy density $\frac{E}{V}$ and the quantum equation of state as

$$\text{Number density}: \frac{N}{V} = \frac{4\pi g}{(2\pi\hbar)^3}\int_0^\infty \frac{p^2}{\exp\left(\frac{\epsilon(p)-\mu}{k_BT}\right) \mp 1}\,dp, \qquad (9.2.1)$$

$$\text{Energy density}: \frac{E}{V} = \frac{4\pi g}{(2\pi\hbar)^3}\int_0^\infty \frac{\epsilon(p)\,p^2}{\exp\left(\frac{\epsilon(p)-\mu}{k_BT}\right) \mp 1}\,dp, \qquad (9.2.2)$$

$$\text{Equation of State}: PV = -\Omega$$

$$= \mp k_BT\frac{4\pi gV}{(2\pi\hbar)^3}\int_0^\infty \log\left[1 \mp \exp\left(\frac{\mu-\epsilon(p)}{k_BT}\right)\right]$$

$$\times p^2 dp. \qquad (9.2.3)$$

In these equations we have omitted the suffixes B.E. and F.D., but shall always remember that the *upper sign refers to B.E. statistics* and the *lower sign refers to F.D. statistics*.

An important function is the **density of state** $\frac{dn(\epsilon)}{d\epsilon} = \frac{4\pi g p^2 V}{(2\pi\hbar)^3}\frac{dp}{d\epsilon}\overline{n(\epsilon)}$, denoting the number of microstates with energy ϵ per unit energy interval. In the general case $\epsilon(p) = \sqrt{p^2c^2 + (mc^2)^2}$, and the density of state is

$$\frac{dn(\epsilon)}{d\epsilon} = \frac{4\pi gV}{(2\pi\hbar c)^3}\frac{\epsilon\sqrt{\epsilon^2 - (mc^2)^2}}{\exp\left(\frac{\epsilon-\mu}{k_BT}\right)\mp 1}. \qquad (9.2.4)$$

For the non-relativistic case $\epsilon(p) = \frac{p^2}{2m}$ and the density of state is

$$\frac{dn(\epsilon)}{d\epsilon} = \frac{2\pi gV}{(2\pi\hbar)^3}(2m)^{\frac{3}{2}}\frac{\sqrt{\epsilon}}{\exp\left(\frac{\epsilon-\mu}{k_BT}\right)\mp 1}. \qquad (9.2.5)$$

The first conclusion we draw from Equation 9.2.1 is about the functional dependence of the chemical potential

$$\mu \equiv \mu\left(\frac{N}{V}, T\right). \tag{9.2.6}$$

Thus for constant N

$$dN \equiv \left(\frac{\partial N}{\partial \mu}\right)_T d\mu + \left(\frac{\partial N}{\partial T}\right)_\mu dT = 0 \tag{9.2.7}$$

will yield

$$\left(\frac{\partial \mu}{\partial T}\right)_N = -\frac{\left(\frac{\partial N}{\partial T}\right)_\mu}{\left(\frac{\partial N}{\partial \mu}\right)_T}. \tag{9.2.8}$$

From Equation 9.2.1 we get

$$\left(\frac{\partial N}{\partial T}\right)_\mu = \frac{4\pi g V}{(2\pi\hbar)^3}\frac{1}{k_B T^2}\int_0^\infty \frac{p^2\left(\epsilon\left(p\right)-\mu\right)\exp\left(\frac{\epsilon(p)-\mu}{k_B T}\right)}{\left[\exp\left(\frac{\epsilon(p)-\mu}{k_B T}\right)\mp 1\right]^2}dp, \tag{9.2.9}$$

$$\left(\frac{\partial N}{\partial \mu}\right)_T = \frac{4\pi g V}{(2\pi\hbar)^3}\frac{1}{k_B T}\int_0^\infty \frac{p^2\exp\left(\frac{\epsilon(p)-\mu}{k_B T}\right)}{\left[\exp\left(\frac{\epsilon(p)-\mu}{k_B T}\right)\mp 1\right]^2}dp \tag{9.2.10}$$

$$> 0. \tag{9.2.11}$$

For BE statistics the right hand side of Equation 9.2.9 is positive because of Equation 9.1.5, so that for this statistics

$$\left(\frac{\partial \mu}{\partial T}\right)_N \leq 0 \tag{9.2.12}$$

and the chemical potential μ is a non-increasing function of temperature T.

For FD statistics nothing can yet be said about the sign of $\left(\frac{\partial \mu}{\partial T}\right)_N$.

In order to progress any further with quantum statistics we need the energy spectrum $\epsilon\left(p\right)$. We study two cases separately.

(i) **Non-relativistic case:**

For particles of mass m we have $\epsilon(p) = \frac{p^2}{2m}$ and Equations 9.2.1, 9.2.2 and 9.2.3 reduce to

$$\frac{N}{V} = \frac{2\pi g}{(2\pi\hbar)^3} (2m)^{\frac{3}{2}} \int_0^\infty \frac{\epsilon^{\frac{1}{2}}}{\exp\left(\frac{\epsilon-\mu}{k_B T}\right) \mp 1} d\epsilon \tag{9.2.13}$$

$$= \frac{2\pi g}{(2\pi\hbar)^3} (2mk_B T)^{\frac{3}{2}} I_\mp\left(\frac{1}{2}, \mu\right), \tag{9.2.14}$$

$$\frac{E}{V} = \frac{2\pi g}{(2\pi\hbar)^3} (2m)^{\frac{3}{2}} \int_0^\infty \frac{\epsilon^{\frac{3}{2}}}{\exp\left(\frac{\epsilon-\mu}{k_B T}\right) \mp 1} d\epsilon \tag{9.2.15}$$

$$= \frac{2\pi g}{(2\pi\hbar)^3} (2mk_B T)^{\frac{3}{2}} (k_B T) I_\mp\left(\frac{3}{2}, \mu\right), \tag{9.2.16}$$

$$P = \mp k_B T \frac{2\pi g}{(2\pi\hbar)^3} (2m)^{\frac{3}{2}} \int_0^\infty \log\left[1 \mp \exp\left(\frac{\mu-\epsilon}{k_B T}\right)\right] \epsilon^{\frac{1}{2}} d\epsilon. \tag{9.2.17}$$

In Equations 9.2.14 and 9.2.16 we have used the integrals given in Equation 14.9.13 of § 14.

Equation 9.2.17 can be integrated by parts to give us the expression for pressure of a non-relativistic ideal quantum gas

$$P = \frac{2}{3} \frac{E}{V}. \tag{9.2.18}$$

Equation 9.2.18 is valid for both BE and FD systems.

We now proceed to derive the equation of state for quantum systems. We assume that the temperature, though low, is not *very* low so that we can take $\exp\left(\frac{\mu}{k_B T}\right) \ll 1$. Thus

$$\frac{N}{V} = \frac{2\pi g}{(2\pi\hbar)^3} (2mk_B T)^{\frac{3}{2}} \Gamma\left(\frac{3}{2}\right) \sum_{n=1}^\infty (\pm 1)^{n-1} \frac{e^{n\mu/(k_B T)}}{n^{\frac{3}{2}}}$$

$$\approx \frac{2\pi g}{(2\pi\hbar)^3} (2mk_B T)^{\frac{3}{2}} \Gamma\left(\frac{3}{2}\right) e^{\mu/k_B T} \left[1 \pm \frac{e^{\mu/k_B T}}{2^{3/2}}\right], \tag{9.2.19}$$

$$\frac{E}{V} = \frac{2\pi g}{(2\pi\hbar)^3} (2mk_B T)^{\frac{3}{2}} (k_B T) \Gamma\left(\frac{5}{2}\right) \sum_{n=1}^\infty (\pm 1)^{n-1} \frac{e^{n\mu/(k_B T)}}{n^{\frac{5}{2}}}$$

$$\approx \frac{2\pi g}{(2\pi\hbar)^3} (2mk_B T)^{\frac{3}{2}} (k_B T) \Gamma\left(\frac{5}{2}\right) e^{\mu/k_B T} \left[1 \pm \frac{e^{\mu/k_B T}}{2^{5/2}}\right]. \tag{9.2.20}$$

With the help of Equations 9.2.18, 9.2.19 and 9.2.20 we obtain the **Virial Equation of State** for non-interacting quantum systems

$$PV = Nk_BT \left[1 + B(T)\frac{N}{V} + \cdots \right], \tag{9.2.21}$$

with the **Second Virial Coefficient**

$$B(T) = \mp \frac{1}{2g} \left(\frac{\pi \hbar^2}{mk_BT} \right)^{3/2}. \tag{9.2.22}$$

It should be noted that even though the system consists of non-interacting particles, quantum nature of the statistics itself introduces a correlation, called **quantum correlation**, and this quantum correlation operates in opposite direction for BE and FD statistics.

(ii) **Mass zero particles and Extreme Relativistic case:**

The relativistic expression for energy $\epsilon(p) = \sqrt{(pc)^2 + (mc^2)^2}$ reduces to $\epsilon(p) = pc$ if the rest mass $m = 0$ or for an extreme relativistic particle for which $pc \gg mc^2$. The most important *mass zero particles* of physics are *photons* and *gravitons* which are *bosons* and the *neutrino* which is a *fermion*.

For these types of energy spectrum the expressions for energy density $\frac{E}{V}$ and the pressure P given by Equations 9.2.2 and 9.2.3 reduce to

Energy density :
$$\frac{E}{V} = \frac{4\pi g}{(2\pi\hbar)^3} \int_0^\infty \frac{\epsilon(p)\,p^2}{\exp\left(\frac{\epsilon(p)-\mu}{k_BT}\right) \mp 1} dp$$

$$= \frac{4\pi g}{(2\pi\hbar)^3} \frac{1}{c^3} \int_0^\infty \frac{\epsilon^3}{\exp\left(\frac{\epsilon-\mu}{k_BT}\right) \mp 1} d\epsilon, \tag{9.2.23}$$

Pressure :
$$P = \mp k_BT \frac{4\pi g}{(2\pi\hbar)^3} \int_0^\infty \log\left[1 \mp \exp\left(\frac{\mu - \epsilon(p)}{k_BT} \right) \right] p^2 dp$$

$$= \mp \frac{k_BT}{c^3} \frac{4\pi g}{(2\pi\hbar)^3} \int_0^\infty \log\left[1 \mp \exp\left(\frac{\mu - \epsilon}{k_BT} \right) \right] \epsilon^2 d\epsilon$$

$$= \frac{1}{3} \frac{4\pi g}{(2\pi\hbar)^3} \frac{1}{c^3} \int_0^\infty \frac{\epsilon^3}{\exp\left(\frac{\epsilon-\mu}{k_BT}\right) \mp 1} d\epsilon$$

$$= \frac{1}{3}\frac{E}{V}. \tag{9.2.24}$$

The similar expression for **radiation pressure** that we know from thermodynamics is thus not special for photon radiation, but is a consequence of the *zero value* of the rest mass and is shared by all such particles, be they bosons or fermions.

9.3   Black Body Radiation and Planck's Law

In the case of **black body radiation**, equilibrium is achieved by emission and absorption of photons by matter. So, total number of photon is not a constant. The condition of thermal equilibrium at a particular temperature T and volume V is given by

$$\left(\frac{\partial F}{\partial N}\right)_{T,V} = 0. \tag{9.3.1}$$

From Equation 4.4.10 we conclude that at equilibrium

$$\mu = 0. \tag{9.3.2}$$

This condition on μ exists for all particles whose number is *not* a constant at thermal equilibrium.

Thus *occupancy* in the $k-$level of energy ϵ_k is given by **Planck's distribution formula**

$$\overline{n_k} = \frac{1}{\exp\left(\frac{\epsilon_k}{k_B T}\right) - 1} \tag{9.3.3}$$

$$= \frac{1}{\exp\left(\frac{\hbar\omega_k}{k_B T}\right) - 1}, \tag{9.3.4}$$

where we have used Planck's *photon hypothesis* $\epsilon_k = \hbar\omega_k$ and the total energy of the photon system becomes

$$E = \sum_k \epsilon_k \overline{n_k} = \sum_k \hbar\omega_k \overline{n_k}. \tag{9.3.5}$$

For large value of the volume V the summation over the photon modes can be replaced by phase space integration over a continuous frequency spectrum in the form

$$\sum_k f(\omega_k) = \int \frac{4\pi g_{\text{photon}} V \left(\frac{\hbar\omega}{c}\right)^2 d\left(\frac{\hbar\omega}{c}\right)}{(2\pi\hbar)^3} f(\omega). \tag{9.3.6}$$

We have used here the experimental result of Compton scattering that a photon of frequency ω has the momentum $\frac{\hbar\omega}{c}$ and a photon of intrinsic spin $J = 1$ has spin degeneracy $g_{\text{photon}} = 2$ corresponding to the *two* transverse modes of polarization. So the average number dN_ω and energy dE_ω of photons with frequency in the range ω and $\omega + d\omega$ are

$$dN_\omega = \frac{V}{\pi^2 c^3} \frac{\omega^2 d\omega}{\exp\left(\frac{\hbar\omega}{k_B T}\right) - 1}, \tag{9.3.7}$$

$$dE_\omega = \frac{V\hbar}{\pi^2 c^3} \frac{\omega^3 d\omega}{\exp\left(\frac{\hbar\omega}{k_B T}\right) - 1}. \tag{9.3.8}$$

For low frequency, when the photon energy $\hbar\omega \ll k_B T$, the thermal energy, we obtain the **Rayleigh-Jeans' formula**

$$dE_\omega = \frac{k_B T V}{\pi^2 c^3} \omega^2 d\omega; \tag{9.3.9}$$

while for high frequency when the photon energy $\hbar\omega \gg k_B T$, we get the **Wien's formula**

$$dE_\omega = \frac{V\hbar}{\pi^2 c^3} \omega^3 \exp\left(-\frac{\hbar\omega}{k_B T}\right) d\omega. \tag{9.3.10}$$

The other thermodynamic functions of black body radiation are as follows.

(i) *Helmholtz' Free Energy, F:*

$$F_{\text{rad}} = -\frac{4V}{3c}\sigma T^4, \tag{9.3.11}$$

where

$$\sigma = \frac{\pi^2 k_B^4}{60\hbar^3 c^2} \tag{9.3.12}$$

is the **Stefan-Boltzmann constant**.

(ii) *Entropy:*

$$S_{\text{rad}} = -\left(\frac{\partial F_{\text{rad}}}{\partial T}\right)_V = \frac{16V}{3c}\sigma T^3 = -\frac{4F_{\text{rad}}}{T}. \tag{9.3.13}$$

(iii) *Internal Energy:*

$$E_{\text{rad}} = F_{\text{rad}} + TS_{\text{rad}} = \frac{4V}{c}\sigma T^4 = -3F_{\text{rad}}. \tag{9.3.14}$$

(iv) *Heat Capacity at constant volume:*

$$C_v = T\left(\frac{\partial S_{\text{rad}}}{\partial T}\right)_V = \frac{16V}{c}\sigma T^3 = \frac{4E_{\text{rad}}}{T}. \tag{9.3.15}$$

(v) *Pressure*:

$$P_{\text{rad}} = -\left(\frac{\partial F_{\text{rad}}}{\partial V}\right)_V = \frac{4}{3c}\sigma T^4 = \frac{1}{3}\frac{E_{\text{rad}}}{V}. \qquad (9.3.16)$$

Thus we arrive at the same relationship between pressure and energy density as was obtained in Equation 9.2.24.

9.4   Lattice Specific Heat and Phonons

The methods of statistical physics are applied to solids to obtain various thermodynamical quantities. The atoms in a crystalline solid perform small oscillations about their equilibrium positions on lattices. If there are N molecules each with ν atoms, then the system has a total of $3N\nu$ degrees of freedom. If we omit the *three translational* and *three rigid rotational* degrees of freedom then the remaining $3N\nu - 6$ degrees of freedom constitute the vibrational modes of the system. These *normal vibrational modes* of the solid, called **phonons**, are so many linear harmonic oscillators. We *state without proof* the results obtained in Solid State Physics about these vibrational modes, $\omega_\alpha(\mathbf{k})$ corresponding to each value of the *wave vector* $\mathbf{k}$. For *3-dimensional solids* there are *three acoustical modes, one longitudnal* and *two transverse*, whose frequencies go to zero as the magnitude $|\mathbf{k}|$ of the wave vector goes to zero. These are the low frequancy phonons having an upper cut-off. The three states of polarization makes the phonons a system with spin $S = 1$ and thus a system of bosons.

The remaining high frequency modes, called *optical phonons*, are confined within a narrow frequency band. This is schematically shown in Figure 9.4.1.

Using the results of § 4.3.4 we can write for the Free Energy of the phonon modes

$$F = N\epsilon_0 + k_B T \sum_\alpha \sum_\mathbf{k} \log\left[1 - \exp\left(-\frac{\hbar\omega_\alpha(\mathbf{k})}{k_B T}\right)\right]. \qquad (9.4.1)$$

We investigate the contributions of phonons to different thermodynamical functions separately for low and high temperature limits.

(i) *Low temperature limit*: At low temperature only those frequencies for which $\hbar\omega_\alpha(\mathbf{k}) \sim k_B T$ contribute to the Free energy F. So we need considering only the acoustic phonons with frequency ω_{acoust} and wavelength $\lambda_{\text{acoust}} \sim a$, where a is the lattice constant. If $\bar{u}$ is the wave

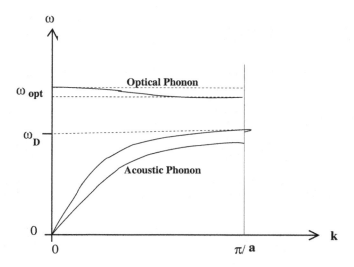

Fig. 9.4.1 Schematic plot of the phonon spectrum of a crystalline solid showing the existence of acoustic and optical phonon modes.

velocity of these phonons then $\omega_{\text{acoust}} = \frac{\bar{u}}{a}$, and the range of temperature for this approximation is $k_B T \ll \frac{\hbar \bar{u}}{a}$.

At such low temperature we can consider the solid as a *continuum*. The number of modes with wave vector between k and $k + dk$ corresponding to frequency ω and $\omega + d\omega$ (taking into account that there is one longitudinal mode with wave velocity u_l and two transverse modes with wave velocity u_t) is given by

$$\frac{4\pi V k^2 dk}{(2\pi)^3} = \frac{V \omega^2 d\omega}{2\pi^2} \left(\frac{1}{u_l^3} + \frac{2}{u_t^3} \right)$$

$$= \frac{3V}{2\pi^2 \bar{u}^3} \omega^2 d\omega, \tag{9.4.2}$$

where we have defined $\bar{u}$ as

$$\frac{3}{\bar{u}^3} = \left(\frac{1}{u_l^3} + \frac{2}{u_t^3} \right). \tag{9.4.3}$$

An upper cut-off ω_D of the frequency known as the **Debye frequency** is required by the physical requirement that the total number of modes is finite.

$$3N\nu = \int_0^{\omega_D} \frac{3V}{2\pi^2\bar{u}^3}\omega^2 d\omega = \frac{V}{2\pi^2\bar{u}^3}\omega_D{}^3. \qquad (9.4.4)$$

In this *continuum model* the Free energy takes the form

$$F = N\epsilon_0 + k_B T \frac{3V}{2\pi^2\bar{u}^3} \int_0^{\omega_D} \omega^2 \log\left[1 - \exp\left(-\frac{\hbar\omega}{k_B T}\right)\right] d\omega. \qquad (9.4.5)$$

Since at the low temperature limit main contribution to the integral on the right hand side comes from small values of ω, so we can as well write

$$F = N\epsilon_0 + k_B T \frac{3V}{2\pi^2\bar{u}^3} \int_0^{\infty} \omega^2 \log\left[1 - \exp\left(-\frac{\hbar\omega}{k_B T}\right)\right] d\omega. \qquad (9.4.6)$$

The integral in Equation 9.4.6 is evaluated by using the results of § 14.9 and the different thermodynamic functions come out in the *continuum model* of Debye as:

(a) *Helmholtz' Free Energy:*

$$F_{\text{phonon}} = N\epsilon_0 - \frac{\pi^2 V k_B^4}{30\left(\hbar\bar{u}\right)^3}T^4. \qquad (9.4.7)$$

(b) *Entropy:*

$$S_{\text{phonon}} = -\left(\frac{\partial F_{\text{phonon}}}{\partial T}\right)_V = \frac{2\pi^2 V k_B^4}{15\left(\hbar\bar{u}\right)^3}T^3. \qquad (9.4.8)$$

(c) *Internal Energy:*

$$E_{\text{phonon}} = F_{\text{phonon}} - TS_{\text{phonon}} = \frac{\pi^2 V k_B^4}{10\left(\hbar\bar{u}\right)^3}T^4. \qquad (9.4.9)$$

(d) *Heat Capacity at Constant Volume:*

$$C_v^{\text{phonon}} = T\left(\frac{\partial S_{\text{phonon}}}{\partial T}\right)_V = \frac{2\pi^2 V k_B^4}{5\left(\hbar\bar{u}\right)^3}T^3. \qquad (9.4.10)$$

Equation 9.4.10 is Debye's celebrated T^3-**law**.

(ii) *High Temperature limit:* Here $k_B T \gg \hbar\omega_\alpha$ and the main contribution to the integral in Equation 9.4.1 comes from the optical phonons with frequencies lying in a narrow band. Since for all these frequencies $1 - \exp\left(-\frac{\hbar\omega_\alpha}{k_B T}\right) \approx \frac{\hbar\omega_\alpha}{k_B T}$, Equation 9.4.1 reduces to

$$F_{\text{phonon}} = N\epsilon_0 + k_B T \sum_\alpha \log\left(\frac{\hbar\omega_\alpha}{k_B T}\right)$$

$$= N\epsilon_0 + 3N\nu k_B T \log\left(\frac{\hbar\bar\omega}{k_B T}\right), \qquad (9.4.11)$$

where we have defined an average frequency $\bar\omega$ as

$$\log\left(\frac{\hbar\bar\omega}{k_B T}\right) = \frac{1}{3N\nu} \sum_\alpha \log\left(\frac{\hbar\omega_\alpha}{k_B T}\right). \qquad (9.4.12)$$

The internal energy become

$$\text{Internal Energy}: E_{\text{phonon}} = F_{\text{phonon}} + T S_{\text{phonon}}$$

$$= N\epsilon_0 + 3N\nu k_B T, \qquad (9.4.13)$$

consistent with the classical *equipartition of energy*, and the classical **Dulong-Petit's law**

$$C_v^{\text{phonon}} = 3N\nu k_B \qquad (9.4.14)$$

follows.

9.5 Degenerate Bose Gas, Bose Condensation

We have already observed in Equations 9.1.6 and 9.2.12 that for finite-mass particles obeying BE statistics the chemical potential μ is a non-positive and non-increasing function of temperature T. So, as the temperature falls μ increases and at a temperature T_0 becomes 0. Thereafter μ remains zero even as the temperature falls further, as we have shown in Figure 9.5.1. Inserting the value $\mu = 0$ and $T = T_0$ in Equation 9.2.14 we get for the number density

$$\frac{N}{V} = \frac{2\pi g}{(2\pi\hbar)^3} (2mk_B T_0)^{\frac{3}{2}} I_-\left(\frac{1}{2}, \mu = 0\right)$$

$$= \frac{2\pi g}{(2\pi\hbar)^3} (2mk_B T_0)^{\frac{3}{2}} \Gamma\left(\frac{3}{2}\right) \zeta\left(\frac{3}{2}\right). \qquad (9.5.1)$$

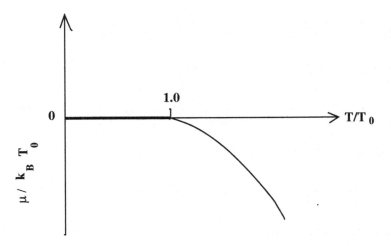

Fig. 9.5.1   Variation of Chemical Potential, μ, as a function of Temperature, T, for Bose Distribution.

Here we have used Equation 14.9.1 for the definition of Riemannian ζ-function. Equation 9.5.1 gives us the value of T_0 for a given value of the particle density $\frac{N}{V}$ of a system of bose particles of mass m.

For temperature $T < T_0$ we *cannot* use Equation 9.2.14; since then the number of particle N in a fixed volume V would seem to depend on temperature. To understand the physical scenario we consider Equation 9.2.16 for the energy of the system for $T < T_0$

$$
\begin{aligned}
E &= \frac{2\pi gV}{(2\pi\hbar)^3} \, (2mk_BT)^{\frac{3}{2}} \, (k_BT) \, I_- \left(\frac{3}{2}, \mu = 0\right) \\
&= \frac{2\pi gV}{(2\pi\hbar)^3} \, (2mk_BT)^{\frac{3}{2}} \, (k_BT) \, \Gamma\left(\frac{5}{2}\right) \zeta\left(\frac{5}{2}\right) \\
&= \frac{3}{2} Nk_BT \frac{\zeta\left(\frac{5}{2}\right)}{\zeta\left(\frac{3}{2}\right)} \left(\frac{T}{T_0}\right)^{\frac{3}{2}} \propto T^{\frac{5}{2}}.
\end{aligned} \tag{9.5.2}
$$

This temperature dependence of energy E shows that particles are transferred to zero-energy state as temperature is lowered below T_0, starting what is known as **Bose Condensation**. T_0, defined by Equation 9.5.1, is known as **Bose Condensation Temperature** and a Bose system below T_0 is called a **Bose Condensate**.

The expression obtained from Equation 9.2.14 for temperature $T < T_0$

$$N_{\epsilon>0} = \frac{2\pi gV}{(2\pi\hbar)^3} (2mk_BT)^{\frac{3}{2}} I_- \left(\frac{1}{2}, \mu = 0\right)$$

$$= \frac{2\pi gV}{(2\pi\hbar)^3} (2mk_BT)^{\frac{3}{2}} \Gamma\left(\frac{3}{2}\right) \zeta\left(\frac{3}{2}\right)$$

$$= N \left(\frac{T}{T_0}\right)^{\frac{3}{2}} \tag{9.5.3}$$

represents the number of particle with non-zero energy which are in *normal state*. The number of particle in the condensate with zero energy is given by

$$N_{\epsilon=0} = N - N_{\epsilon>0} = N\left[1 - \left(\frac{T}{T_0}\right)^{\frac{3}{2}}\right]. \tag{9.5.4}$$

Other thermodynamic functions for the condensate are

(i) *Heat Capacity at constant volume:*

$$C_v = \left(\frac{\partial E}{\partial T}\right)_V = \frac{5}{2}\frac{E}{T} \propto T^{\frac{3}{2}}, \tag{9.5.5}$$

(ii) *Entropy:*

$$S = \int \frac{C_v}{T} dT = \frac{5}{3}\frac{E}{T} \propto T^{\frac{3}{2}}, \tag{9.5.6}$$

(iii) *Helmholtz' Free Energy:*

$$F = E - TS = -\frac{2}{3}E \propto T^{\frac{5}{2}}, \tag{9.5.7}$$

(iv) *Gibbs' Potential*

$$G = N\mu = 0, \tag{9.5.8}$$

(v) *Pressure:*

$$P = \frac{2}{3}\frac{E}{V} \propto T^{\frac{5}{2}}. \tag{9.5.9}$$

In Figure 9.5.2 we have sketched the dependence of Heat Capacity of the Bose Condensate as a function of temperature showing clearly the *cusp* at $T = T_0$.

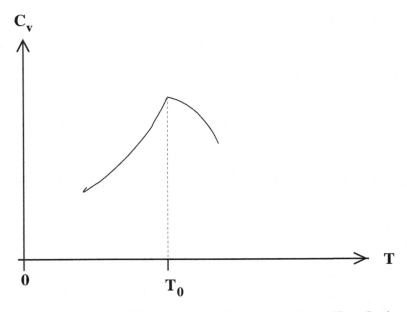

Fig. 9.5.2 Schematic plot of Heat Capacity C_v at constant volume of Bose Condensate as a function of temperature T.

9.6 Liquid He and Superfluidity

9.6.1 *Systematics of Liquid ^4He*

Equation 9.5.1 for the Bose condenstation temperature T_0 can be written as

$$T_0 \;=\; \frac{2\pi\hbar^2}{k_B} \left(\frac{N_A}{M}\right)^{\frac{5}{3}} \left(\frac{\rho}{\zeta\left(\frac{3}{2}\right)}\right)^{\frac{2}{3}}, \tag{9.6.1}$$

in terms of the mass density ρ and Molecular Weight M of the system and the fundamental constants Avogadro Constant $N_A = 6.022 \times 10^{23}$, Boltzmann Constant $k_B = 1.38 \times 10^{-16}$ erg/K, Planck's Constant $\hbar = 1.054 \times 10^{-27}$ erg $-$ sec and $\zeta\left(\frac{3}{2}\right) = 2.612\cdots$. Thus the lower the molecular weight of the boson system the higher will be T_0. Since among the elements of the Periodic Table ^4He with $M = 4.002602$ is the lightest stable bosonic system, it is expected that it will undergo Bose condensation at a suitably high temperature. Using liquid Helium's density $\rho = 124.8 \times 10^{-3}$ g/cm^3 at its melting point, we get $T_0 = 2.822\ K$.

In Figure 9.6.1 we have plotted the very interesting phase diagram for ^4He. When temperature is lowered even up to absolute zero ^4He does not

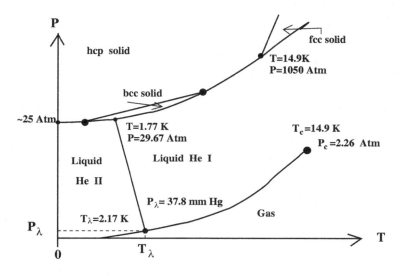

Fig. 9.6.1   Schematic plot of the phase diagram of ^4He.

solidify under its own vapor pressure, but only under an external pressure of ~ 25 atmosphere. In solid phase it exists in all the important crystal structures: hexagonal close packed (hcp), face-centered cube (fcc) and body-centered cube (bcc). Very few material has so many varieties of crystal structures. But its unique property appears when liquid ^4He is cooled by pumping out its overlying vapor. The boiling of the liquid which was present so long suddenly stops at a temperature $T_\lambda = 2.17\ K$ and the liquid enters a new phase called $He\ II$. By contrast the normal phase is called $He\ I$. This value of T_λ was so near the calculated value of T_0 for ^4He as to make people conclude that $He\ II$ phase is the Bose Condensation phase.

We first describe the experimentally oberved properties of liquid He in its $He\ II$ phase.

(i) *Infinite Thermal Conductivity.*

First was the already mentioned property of stoppage of bubble formation even when evaporation was vigorously continuing. In ordinary liquids bubble forms during boiling because the localized accumulation of heat fails to be dispersed, as the thermal conductivity is finite. Absence of bubble formation in liquid $He\ II$ during boiling establishes that **thermal conductivity** $\kappa_{\text{therm}} = \infty$. If we compare the thermal conductivity κ_{II} of liquid $He\ II$ with κ_{Cu}, that of one of the most conductive metal, we find within experimental accuracy $\kappa_{II} \sim 800 \times \kappa_{Cu}$. Also if κ_I is the thermal conductivity of liquid He in the normal phase, then $\kappa_{II} \approx 13.5 \times 10^6\ \kappa_I$.

(ii) *Zero Viscosity;*

The second property is the absence of viscosity in $He\ II$ phase, from which the name **Superfluid** has been given to this phase. When liquid He is made to flow through a capillary of diameter $\sim 10^{-4}$ cm flow of He in the normal phase stops, while liquid $He\ II$ with velocity less than a *critical value* v_c flows with almost **zero viscosity** $\eta_{II} = 0$, *i.e.* with zero heat dissipation. In terms of viscosity η_I of the normal phase, $\eta_{II} \approx 10^{-3}\ \eta_I$.

If, however, viscosity is measured by rotating disc method it is found that $\eta \neq 0$ even below T_λ. This made Tisza to propose the *Two-Fluid Model* where liquid He below T_λ is assumed to have two components *normal* and *superfluid* with mass densities ρ_n and ρ_s respectively, the total density ρ being

$$\rho = \rho_n + \rho_s. \tag{9.6.2}$$

(iii) *Zero Surface Tension.*

In the **Double beaker Creeping flow experiment** of superfluid ^4He, schematically shown in Figure 9.6.2, it is found that a film of liquid $He\ II$ of thickness of several atomic layers creeps over the separating wall of a double beaker to find the level of He of the container kept below the beaker. Since finite capillary rise is due to finite value of surface tension this experiment proves that superfluid liquid He has **zero Surface Tension**.

(iv) *Thermomechanical Effects.*

Two chambers A and B containing ^4He in the superfluid phase are connected by a *superleak* as shown in Figure 9.6.3. Initially temperature T and pressure P are the same in both the chambers. As the superfluid flows from chamber A to chamber B the pressure head moves, untill the pressure difference ΔP and the temperature difference ΔT satisfy

$$\Delta P = \rho s \Delta T, \tag{9.6.3}$$

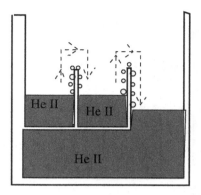

Fig. 9.6.2   Creeping flow of ^4He in the superfluid phase.

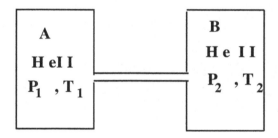

Fig. 9.6.3   Explanation of thermomechanical effects of ^4He in the superfluid phase.

where s is the *specific entropy*, i.e. total entropy per unit mass. This can be understood from the equilibrium condition

$$\Delta G = -S\Delta T + V\Delta P \ = \ 0, \tag{9.6.4}$$

$$\Delta P = \frac{S}{V}\Delta T \ = \ s \cdot \rho \cdot \Delta T. \tag{9.6.5}$$

There are two manifestations of thermomechanical effects which we now describe.

(a) *Mechanocaloric Effect.* In the experimental set-up for observing mechanocaloric effect for superfluid ^4He, sketched in Figure 9.6.4, the capillary opening of the upper beaker A is filled with emery powder with space between them of linear dimension 10^{-3} cm so that *only* superfluid component flows to the lower container B. It is

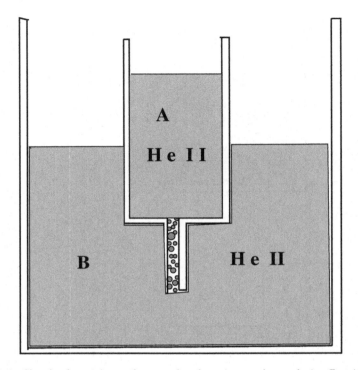

Fig. 9.6.4   Sketch of experimental set-up for observing mechanocaloric effect in super-fluid ^4He.

experimentally observed that temperature of the container A rises, thus establishing that superfluid component carries **zero entropy**.

(b) *Fountain Effect.* The fountain effect in superfluid ^4He, shown in Figure 9.6.5 is the reverse experimental set-up. Here a temperature difference produces a pressure difference to move the superfluid. The U-tube is packed with 1 μ sized emery powder. As a narrow light beam is flashed at the point A, superfluid flows through the capillary from A to B and forms a fountain at the end B, due to the resulting pressure difference.

(v) *Specific Heat.*

In Figure 9.6.6 we have drawn a schematic plot of the heat capacity of ^4He as a function of temperature. At the temperature T_λ specific heat shows a singularity. The similarity of the graph with the letter λ of the Greek alphabet has given the name λ transition and the λ-point T_λ.

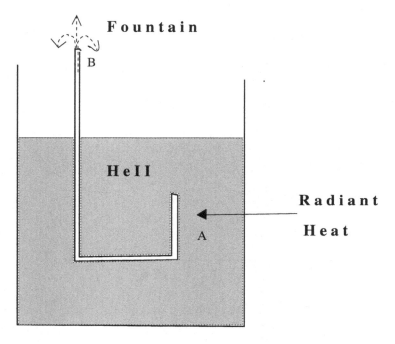

Fig. 9.6.5   Fountain effect in superfluid ^4He.

This figure has an important bearing on the nature of the λ−transition. If we compare Figure 9.6.6 with Figure 9.5.2 their difference becomes obvious. The λ-transition of $He\ II$ cannot thus be wholly a Bose condensation. However, people are convinced that there is an important Bose condensation phase present in liquid $He\ II$.

9.6.2   *Landau's 2-Fluid Model*

Landau developed the phenomenological theory of Tisza in a more profound way. According to Landau liquid $He\ II$ is considered to be consisting of *a perfect background fluid with zero entropy and viscosity* and *phonon excitations* like the phonon excitations in a solid lattice. The phonon excitations, according to Landau, are the normal component of the fluid. At low temperature specific heat of the phonon gas is $\propto T^3$. As the perfect background flows through a superleak, the phonons are hindered because they collide with the boundary wall. The two-fluid model is further exhibited by the ingeneous experiment due to Andronikashvili. In this experiment, a pile of

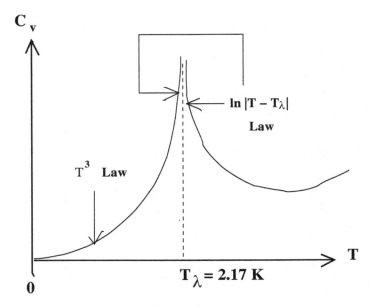

Fig. 9.6.6   Schematic plot of the heat capacity of ^4He as a function of temperature T.

closely packed discs is rotated in a liquid *He II* bath. The rotation leaves
the superfluid component unaffected, but the phonons and other excitations are dragged around with the discs and have an inertial effect (giving
rise to finite viscosity) which can be measured.

The two components can oscillate in opposite phase in such a way that
the total mass density $\rho = \rho_n + \rho_s$ of *He II* remains constant. The density
of excitations is a function of temperature and the temperature dependence
of the fraction of normal component in liquid ^4He is shown in Figure 9.6.7.

This leads to a new type of wave propagation, known as **second sound**,
which is a temperature wave and is excited by heat rather than pressure
pulses as in ordinary sound (called *first sound*) propagation. According to
Landau theory the speed of the second sound is $c_s = \frac{u}{\sqrt{3}}$, where u is the
speed of the first sound. Actual observation, however, shows that at very
low temperature c_s exceeds $\frac{u}{\sqrt{3}}$ as is shown in Figure 9.6.8.

To explain the region of the curve between T_λ and 1 K Landau introduced a second type of excitations **rotons** along with phonons. Landau
also introduced (which was later theoretically established) the dispersion
spectrum of ^4He showing *phonon* and *roton*, sketched in Figure 9.6.9.

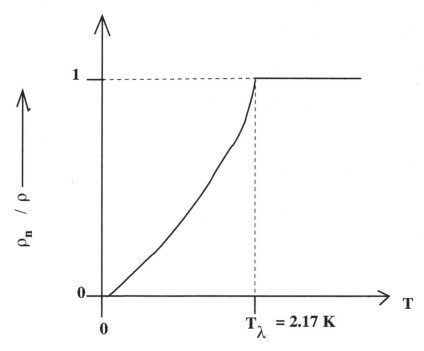

Fig. 9.6.7   Temperature dependence of the fraction of normal component in liquid ^4He.

The dispersion relations for phonons and rotons are

$$\text{Phonons}: \epsilon(p) = u \cdot p, \tag{9.6.6}$$

$$\text{Rotons}: \epsilon(p) = \Delta + \frac{(p - p_0)^2}{2m^*}, \tag{9.6.7}$$

with the best-fit values $\Delta = 8.9\ K$, $p_0 = 2.1 \times 10^{-19}$ g-cm/s. The best-fit value for the effective mass $m^* = 1.72 \times 10^{-24}$ g is comparable with that of Hydrogen and not that of Helium. Thus the Helium atoms themselves *cannot* be identified with the **quasi-particles** rotons.

Once knowing the dispersion curve, Equations 9.2.1 and 9.2.3 can be used to derive the thermodynamic quantities for the quasi-particles (since

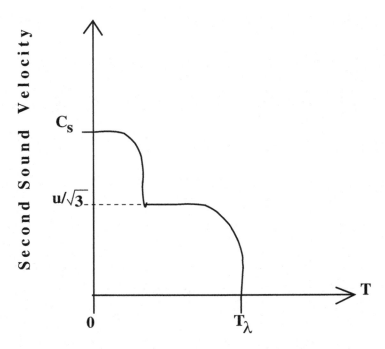

Fig. 9.6.8   Temperature dependence of sound velocity in liquid ^4He.

at a temperature below T_λ, $\mu = 0$ and for ^4He the spin degeracy $g = 1$)

$$\text{Number}: N_{\text{qp}} = \frac{V}{(2\pi\hbar)^3} \int_0^\infty \frac{d^3\mathbf{p}}{\exp\left(\frac{\epsilon(p)}{k_B T}\right) - 1}, \qquad (9.6.8)$$

$$\text{Free Energy}: F_{\text{qp}} = \Omega_{\text{qp}} = \frac{V k_B T}{(2\pi\hbar)^3} \int_0^\infty \log\left[1 - \exp\left(-\frac{\epsilon(p)}{k_B T}\right)\right] d^3\mathbf{p}.$$
$$(9.6.9)$$

The thermodynamic relations for entropy $S = -\left(\frac{\partial F}{\partial T}\right)_V$ and heat capacity at constant volume $C_v = T\left(\frac{\partial S}{\partial T}\right)_v$ will finally give us for *phonon gas*

$$N_{\text{ph}} = \zeta(3)V\left(\frac{k_B T}{\hbar u}\right)^3, \qquad (9.6.10)$$

$$F_{\text{ph}} = -\frac{\pi^2}{90}V k_B T\left(\frac{k_B T}{\hbar u}\right)^3, \qquad (9.6.11)$$

$$C_v^{\text{ph}} = \frac{2\pi^2}{15}V\left(\frac{k_B T}{\hbar u}\right)^3 \quad \propto T^3. \qquad (9.6.12)$$

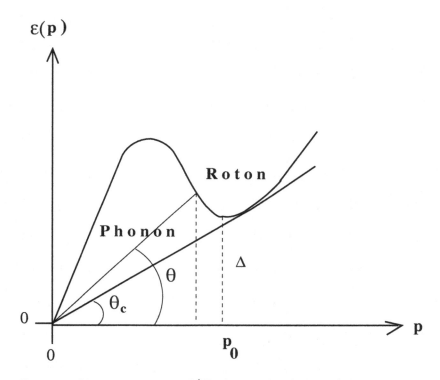

Fig. 9.6.9   Dispersion spectrum of ^4He showing the *phonon* and the *roton* part.

and for the roton gas

$$N_{\text{rot}} = \frac{2Vp_0^2}{(2\pi\hbar^2)^{\frac{3}{2}}} (m^* k_B T)^{\frac{1}{2}} \exp\left(-\frac{\Delta}{k_B T}\right), \tag{9.6.13}$$

$$F_{\text{rot}} = -k_B T N_{\text{rot}}, \tag{9.6.14}$$

$$C_v^{\text{rot}} \propto \left[\frac{3}{4} + \frac{\Delta}{k_B T} + \left(\frac{\Delta}{k_B T}\right)^2\right] \exp\left(-\frac{\Delta}{k_B T}\right). \tag{9.6.15}$$

At temperature below 0.8 K main contribution to heat capacity comes from phonons.

We now evaluate the effective mass parameter m^* and the mass density ρ_n of the normal component. We imagine that the gas of quasi-particles moves as a whole with respect to the liquid with velocity **v**. Thus the liquid is moving with respect to the gas with velocity -**v**. In the rest frame of the gas of excitations, we have

$$\text{Energy of the elementary excitation} = \epsilon - \mathbf{p} \cdot \mathbf{v}, \tag{9.6.16}$$

where ϵ is the energy of the excitation when the liquid is at rest. Then the average momentum of the excitation gas is given by

$$
\begin{aligned}
\langle \mathbf{p} \rangle &= \frac{V}{(2\pi\hbar)^3} \int \mathbf{p}\bar{n} \left(\epsilon - \mathbf{p} \cdot \mathbf{v}\right) d^3\mathbf{p} \\
&\approx \frac{V}{(2\pi\hbar)^3} \int \mathbf{p} \left(\mathbf{p} \cdot \mathbf{v}\right) \left(-\frac{\partial \bar{n}}{\partial \epsilon}\right) d^3\mathbf{p} \\
&= \mathbf{v} \cdot \frac{1}{3} \frac{V}{(2\pi\hbar)^3} \int p^2 \left(-\frac{\partial \bar{n}}{\partial \epsilon}\right) d^3\mathbf{p},
\end{aligned}
\tag{9.6.17}
$$

where we have assumed isotropy of the occupancy $\bar{n}$. Equating $\langle \mathbf{p} \rangle = m^* \mathbf{v}$ we get for the effective mass m^* and the mass density $\rho_n = \frac{m^*}{V}$ of the normal component as

$$m^* = \frac{1}{3} \frac{V}{(2\pi\hbar)^3} \int p^2 \left(-\frac{\partial \bar{n}}{\partial \epsilon}\right) d^3\mathbf{p}, \tag{9.6.18}$$

$$\rho_n = \frac{1}{3} \frac{1}{(2\pi\hbar)^3} \int p^2 \left(-\frac{\partial \bar{n}}{\partial \epsilon}\right) d^3\mathbf{p}. \tag{9.6.19}$$

Using the energy spectrum Equations 9.6.6 and 9.6.7 for the two types of excitations we obtain

$$\rho_n = \begin{cases} \frac{2\pi\hbar}{45u} \left(\frac{k_B T}{\hbar u}\right)^4 & \text{for phonons,} \\ \frac{2}{3} \frac{p_0^4}{(2\pi\hbar^2)^{\frac{3}{2}}} \sqrt{\frac{m^*}{k_B T}} \exp\left(-\frac{\Delta}{k_B T}\right) & \text{for rotons.} \end{cases} \tag{9.6.20}$$

Landau's model also establishes the existence of the critical velocity v_c for the superfluid and graphically calculates its value. When a normal liquid flows through a capillary with constant velocity $\mathbf{v}$, there will be friction with the walls of the tube and also within the liquid itself, due to the presence of viscosity. There will be energy dissipation and the flow will gradually slow down. Studying the flow in a coordinate frame moving with the liquid, we note that the helium mass will be stationary in this coordinate system and the capillary wall moves with a velocity -$\mathbf{v}$. If there is viscocity present, its presence will be signified by the gradual appearence of quasi-particle excitations like phonons and rotons in the liquid.

Let us assume that only one quasi-particle with energy $\epsilon(p)$ and momentum $\mathbf{p}$ appears in the liquid. Then in the coordinate frame in which the liquid was initially at rest, the system will have energy equal to $E_0 = \epsilon(p)$ and momentum $\mathbf{P}_0 = \mathbf{p}$. Transforming to the coordinate system in which the capillary is at rest, energy E and momentum $\mathbf{P}$ of the liquid will be given by Galilean transformation

$$E = E_0 - \mathbf{P}_0 \cdot \mathbf{v} + \frac{1}{2}Mv^2$$

$$= \epsilon(p) - \mathbf{p} \cdot \mathbf{v} + \frac{1}{2}Mv^2, \tag{9.6.21}$$

$$\mathbf{P} = \mathbf{P}_0 - M\mathbf{v}. \tag{9.6.22}$$

The last term $\frac{1}{2}Mv^2$ is the kinetic energy of the liquid as a whole and $\epsilon(p) - \mathbf{p} \cdot \mathbf{v}$ is the change of energy of the system because of the appearence of elementary excitation and **dissipation of energy**. Thus the condition for *appearence of viscosity* is

$$\epsilon(p) - \mathbf{p} \cdot \mathbf{v} < 0, \tag{9.6.23}$$

$$|\mathbf{v}| \geq \frac{\epsilon(p)}{|\mathbf{p}|}. \tag{9.6.24}$$

Referring to Figure 9.6.9 we find that the minimum value of $\frac{\epsilon(p)}{|\mathbf{p}|} = \tan\theta$ is $\tan\theta_c$. Thus when the speed of the liquid $v > v_c = \tan\theta_c$, viscosity in the forms of phonons and rotons appear in the liquid.

9.6.3 *Systematics of Liquid* ^3He

There is a fermionic isotope ^3He with relative natural abundance of $1.4 \times 10^{-4}\%$. In Figure 9.6.10 we have shown the phase diagram of the mixture of ^4He and ^3He. This fermionic system is *not* expected to show superfluidity. In Figure 9.6.11 we have drawn the phase diagram of ^3He. At extremely low temperature $\sim mK$ the system has shown transitions to different types of superfluid phases. The phase diagram of ^3He close to its transition to the superfluid phase at this extremely low temperature range has been drawn in Figure 9.6.12. The heat capacity result depicted in Figure 9.6.13 shows that above the transition temperature the normal phase is a degenerate Fermi gas, whose property we shall discuss in § 9.7. There is also no singularity, but only a discontinuity at the transition twmperature. Discussion on the nature of superfluid phase of ^3He, however, goes beyond the domain of Statistical Physics.

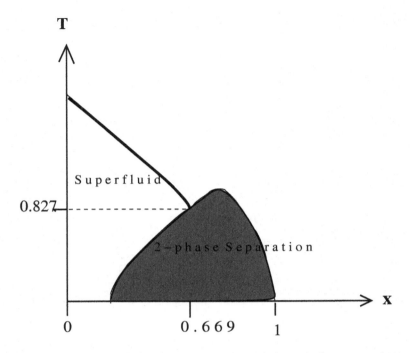

Fig. 9.6.10   Phase diagram of mixture of ^4He and ^3He.

9.7   Degenerate Fermi Gas, Degeneracy Pressure, Specific Heat

We now investigate a non-relativistic fermion system of spin degeneracy g, mass m and energy $\epsilon(p) = \frac{p^2}{2m}$ in the very low temperature region when the thermal energy $k_B T \ll \mu$, the chemical potential. This system is called a **degenerate Fermi gas** and at $T = 0$ it is said to be completely degenerate. Chemical Potential μ is a function of temperature T and its value μ_0 for $T = 0$ K is called the **Fermi Energy** and is denoted by ϵ_F.

At $T = 0K$ the occupancy

$$\overline{n}(\epsilon) = \lim_{T \to 0} \frac{1}{\exp\left(\frac{\epsilon - \mu}{k_B T}\right) + 1} \tag{9.7.1}$$

is a step function as shown in Figure 9.7.1 All the energy states below $\epsilon_F = \mu_0$ are occupied with occupancy 1 and all those above ϵ_F are empty

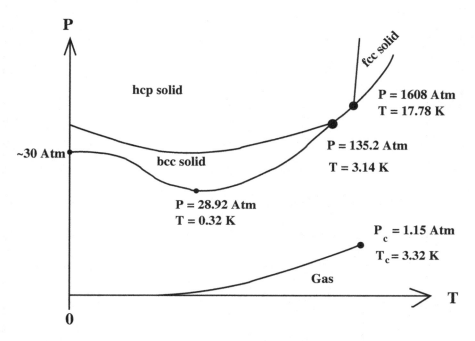

Fig. 9.6.11   Schematic plot of the phase diagram of ^3He.

with occupancy 0. Defining **Fermi Momentum** p_F by the relation

$$\epsilon_F = \frac{p_F^2}{2m}, \tag{9.7.2}$$

we get for the particle density

$$\frac{N}{V} = \frac{4\pi g}{(2\pi\hbar)^3} \int_0^{p_F} p^2 dp = \frac{4\pi g}{3(2\pi\hbar)^3} p_F^3. \tag{9.7.3}$$

We thus have at $T = 0$ K

$$\text{Fermi Momentum}: p_F = \hbar \left(\frac{6\pi^2}{g}\right)^{\frac{1}{3}} \left(\frac{N}{V}\right)^{\frac{1}{3}}, \tag{9.7.4}$$

$$\text{Fermi Energy}: \epsilon_F = \frac{\hbar^2}{2m} \left(\frac{6\pi^2}{g}\right)^{\frac{2}{3}} \left(\frac{N}{V}\right)^{\frac{2}{3}}, \tag{9.7.5}$$

$$\text{Energy}: E_0 = \frac{4\pi g V}{(2\pi\hbar)^3} \int_0^{p_F} \epsilon(p) p^2 dp = \frac{4\pi g V}{(2\pi\hbar)^3} \frac{1}{2m} \frac{p_F^5}{5}$$

$$= \frac{3}{5} N \epsilon_F. \tag{9.7.6}$$

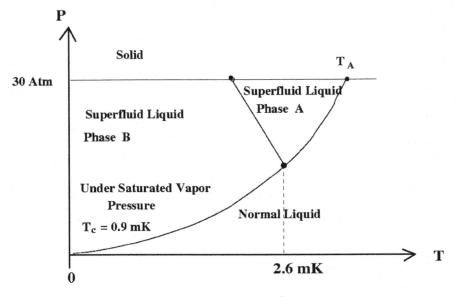

Fig. 9.6.12 Schematic plot of the phase diagram of ^3He near to its transition to the superfluid phase.

The resulting pressure

$$P = \frac{2}{3}\frac{E_0}{V},$$ (9.7.7)

known as the **degeneracy pressure** has *inter alia* immense astrophysical significance. Within the stellar core it is this degeneracy pressure due to the nucleons that balances the overlying gravitational pressure giving the stars their stability. In terms of the Fermi Energy ϵ_F we can write the density of state at any energy as

$$\frac{dn\,(\epsilon)}{d\epsilon} = \frac{3}{2}\frac{N}{\epsilon_F}\frac{\sqrt{\frac{\epsilon}{\epsilon_F}}}{\exp\left(\frac{\epsilon-\mu}{k_BT}\right)+1}.$$ (9.7.8)

For non-zero temperature $T \neq 0$ we can evaluate the expressions Equations 9.2.13 and 9.2.15 for the number and the energy densities with help of the Equations 14.9.15, 14.9.16 and 14.9.17 and the value of $\zeta(2)$ from

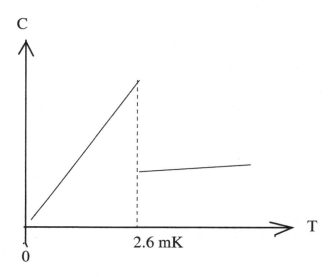

Fig. 9.6.13   Schematic plot of the heat capacity of ^3He.

Equation 14.9.12

$$\frac{N}{V} = \frac{2\pi g}{(2\pi\hbar)^3} (2m)^{\frac{3}{2}} \int_0^\infty \frac{\epsilon^{\frac{1}{2}}}{\exp\left(\frac{\epsilon-\mu}{k_BT}\right)+1} \, d\epsilon$$

$$= \frac{2\pi g}{(2\pi\hbar)^3} (2m)^{\frac{3}{2}} \left[\int_0^\mu \epsilon^{\frac{1}{2}} d\epsilon + 2 \sum_{n=0}^\infty (k_BT)^{2n+2} \left(\left(\frac{d}{d\mu}\right)^{2n+1} \mu^{\frac{1}{2}} \right) \right.$$
$$\left. \times \left(1 - \frac{1}{2^{2n+1}}\right) \zeta(2n+2) \right]$$

$$= \frac{2\pi g}{(2\pi\hbar)^3} (2m)^{\frac{3}{2}} \frac{2}{3} \mu^{\frac{3}{2}} \left[1 + \frac{\pi^2}{8} \left(\frac{k_BT}{\mu}\right)^2 + \cdots \right], \qquad (9.7.9)$$

$$\frac{E}{V} = \frac{2\pi g}{(2\pi\hbar)^3} (2m)^{\frac{3}{2}} \int_0^\infty \frac{\epsilon^{\frac{3}{2}}}{\exp\left(\frac{\epsilon-\mu}{k_BT}\right)+1} \, d\epsilon$$

$$= \frac{2\pi g}{(2\pi\hbar)^3} (2m)^{\frac{3}{2}} \left[\int_0^\mu \epsilon^{\frac{3}{2}} d\epsilon + 2 \sum_{n=0}^\infty (k_BT)^{2n+2} \left(\left(\frac{d}{d\mu}\right)^{2n+1} \mu^{\frac{3}{2}} \right) \right.$$
$$\left. \times \left(1 - \frac{1}{2^{2n+1}}\right) \zeta(2n+2) \right]$$

$$= \frac{2\pi g}{(2\pi\hbar)^3} (2m)^{\frac{3}{2}} \frac{2}{5} \mu^{\frac{5}{2}} \left[1 + \frac{5\pi^2}{8} \left(\frac{k_BT}{\mu}\right)^2 + \cdots \right]. \qquad (9.7.10)$$

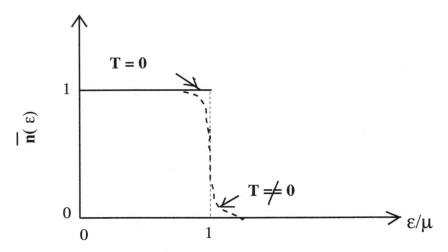

Fig. 9.7.1 Occupancy $\bar{n}(\epsilon)$ for Fermi Statstics at zero degree Kelvin and at non-zero Temperature.

As a series expansion in $\left(\frac{k_B T}{\epsilon_F}\right)^2$ we get

$$\mu = \epsilon_F \left[1 - \frac{\pi^2}{12}\left(\frac{k_B T}{\epsilon_F}\right)^2 + \cdots\right], \tag{9.7.11}$$

$$E = E_0 \left[1 + \frac{5\pi^2}{12}\left(\frac{k_B T}{\epsilon_F}\right)^2 + \cdots\right]. \tag{9.7.12}$$

We thus see that the chemical potential μ for degenerate fermi gas decreases with increasing temperature just like for degenerate bose gas. Another usefull function is the density of state $\frac{dn(\epsilon)}{d\epsilon}$ which gives the number of particle with energy between ϵ and $\epsilon + d\epsilon$. In Figure 9.7.2 we have plotted the fractional density of state $\frac{1}{N}\frac{dn(x)}{dx}$ as a function of $x = \frac{\epsilon}{\mu}$ for two values of temperature $T = 0K$ and $T = 0.05\mu/k_B$. At non-zero temperature particles originally lying in a energy shell $\sim k_B T$ below the fermi energy is transferred in a similar energy shell above the fermi energy.

The heat capacity for free fermion gas has a linear dependence on temperature

$$C_v = \left(\frac{\partial E}{\partial T}\right)_V = N k_B \frac{\pi^2}{2}\left(\frac{k_B T}{\epsilon_F}\right) = \gamma T. \tag{9.7.13}$$

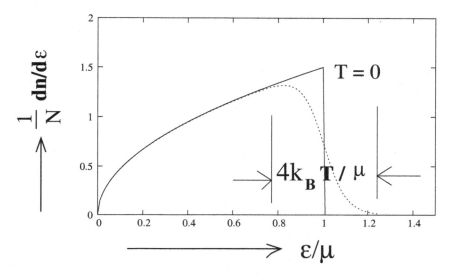

Fig. 9.7.2   Plot of fractional density of state $\frac{1}{N}\frac{dn(x)}{dx}$ as a function $x = \frac{\epsilon}{\mu}$ for Fermi Statstics at zero degree Kelvin (solid curve) and at non-zero Temperature (dashed curve) given by $k_B T = 0.05\ \mu$.

Since ϵ_F varies inversely as the mass of the fermions Heat Capacity varies directly as the fermionic mass involved. In ordinary metals $\gamma \sim \ mJ/K^2$. But in systems like $CeCu_2Si_2$, UAl_3 etc. $\gamma \sim \ J/K^2$ and these systems are called **Heavy Fermion Systems**. This thousand fold increase in the effective mass of the fermion is due to strong interaction among the electrons of the system.

In crystalline solids both this fermionic contribution as also the lattice contribution to heat capacity is present to make the experimental value behave like

$$C_{\text{solid}} = \gamma T + aT^3. \tag{9.7.14}$$

If C_{solid} is plotted against T then we have to draw the tangent to the curve to obtain the value of γ. The error involved is considerable in this procedure. That is why experimentalists plot the graph of $\frac{C_{\text{solid}}}{T}$ against T. The asymptotic value at $T = 0$ gives the value of γ. In this procedure the error introduced in the value of γ is minimal.

9.8   Magnetism of Free Fermions

9.8.1   Preamble

We are considering a system of N Free Fermions of spin S in units of $\hbar$, Landé g–factor g_S, electric charge q in $e.s.u$ and mass m moving in a uniform external magnetic field $\mathbf{B}$, for simplicity taken to be aligned along the z–axis. The system thus has an intrinsic magnetic moment

$$\vec{\mu_0} = g_S \frac{q\hbar}{2mc} \mathbf{S}. \tag{9.8.1}$$

Magnetic moment in this system arises from the orbital motion of the particles as well as the intrinsic magnetic moment of the particles. Niels Bohr and Miss H.J.van Leeuwen proved that in classical statistics the orbital motion does not generate any magnetic moment of the system. L.D.Landau, however, proved the existence of the orbital diamagnetism when quantum statistics is used and this result will be discussed in § 9.8.2. Pauli's proof of the existence of paramagnetism from the second source will be discussed in § 9.8.3.

9.8.2   Landau Diamagnetism

The Hamiltonian for a particle of charge q and mass m in a external magnetic field $\mathbf{B} = \nabla \times \mathbf{A}$ is

$$\mathcal{H} = \frac{1}{2m} \left(\mathbf{p} - \frac{q\mathbf{A}}{c} \right)^2 = \frac{\mathbf{p}^2}{2m} + \frac{q^2\mathbf{A}^2}{2mc^2} - \frac{q}{2mc} \left(\mathbf{p} \cdot \mathbf{A} + \mathbf{A} \cdot \mathbf{p} \right), \tag{9.8.2}$$

where we have used the Gaussian units. We must now remember that in all quantum mechanical operations we are really interested in forms like $\mathcal{H}\psi$ so we can transform

$$\mathbf{p} \cdot (\mathbf{A}\psi) = (\mathbf{p} \cdot \mathbf{A}) \psi + \mathbf{A} \cdot \mathbf{p}\psi. \tag{9.8.3}$$

In defining the magnetic field in terms of the vector potential, we are at a freedom in chosing the value of $\nabla \cdot \mathbf{A}$. In all time-independent cases we generally choose the **Coulomb/London gauge** $\nabla \cdot \mathbf{A} = 0$. In this gauge the hamiltonian will take the form

$$\mathcal{H} = \frac{1}{2m} \left(\mathbf{p} - \frac{q\mathbf{A}}{c} \right)^2 = \frac{\mathbf{p}^2}{2m} + \frac{q^2\mathbf{A}^2}{2mc^2} - \frac{q}{mc} \mathbf{A} \cdot \mathbf{p}. \tag{9.8.4}$$

We now choose the vector potential $\mathbf{A} = yB\hat{\mathbf{x}}$, where $\hat{\mathbf{x}}$ denotes the unit vector along the x-axis, to describe the uniform external magnetic field $\mathbf{B}$ along the positive z−axis, so that $\nabla \times \mathbf{A} = \mathbf{B}$ in the **Coulomb/London gauge** $\nabla \cdot \mathbf{A} = 0$.

The single-particle hamiltonian is now of the form

$$\mathcal{H} = \frac{1}{2m}\left(p_x^2 + p_y^2 + p_z^2\right) + \frac{q^2 B^2}{2mc^2}y^2 + \frac{qB}{mc}yp_x. \tag{9.8.5}$$

Since $[p_x, \mathcal{H}] = 0$, we can treat p_x as a classical number and write

$$\mathcal{H} = \frac{p_z^2}{2m} + \left[\frac{p_y^2}{2m} + \frac{1}{2}m\omega_c^2\left(y + \frac{cp_x}{qB}\right)^2\right], \tag{9.8.6}$$

where

$$\omega_c = \frac{qB}{mc}, \tag{9.8.7}$$

the cyclotron frequency is twice the *Larmor Frequency* $\omega_L = \frac{qB}{2mc}$. The motion of the system is like that of free particles along z−axis and a simple harmonic motion with frequency ω_c along the y−axis with the mean position $-\frac{cp_x}{qB}$. The energy of the system with the two quantum numbers p_z, n is

$$\epsilon_{p_z, n} = \frac{p_z^2}{2m} + \hbar\omega_c\left(n + \frac{1}{2}\right). \tag{9.8.8}$$

These quantized discrete levels are called the **Landau levels**.

We now investigate what the form the phase space integral will take. We note that the harmonic oscillator is centered about a point $-\frac{cp_x}{qB}$ on the y−axis. For a finite sample of linear dimensions L_x, L_y and L_z the position of the origin can have the uncertainy L_y. Thus the uncertainty $\Delta p_x = L_y \frac{qB}{c}$ and

$$\frac{L_x \Delta p_x}{(2\pi\hbar)} = \frac{L_x L_y m\omega_c}{(2\pi\hbar)}. \tag{9.8.9}$$

For the system with spin degeneracy g_S the phase-space integral will now have the form

$$\int \frac{\Delta\Gamma}{(2\pi\hbar)^3}f(\cdots) = \frac{V m g_S \omega_c}{(2\pi\hbar)^2}\sum_n \int dp_z f(\cdots). \tag{9.8.10}$$

There is, of course, restrictions on the limits of the integration and the summation. For zero-temperature case we have

$$0 \le \epsilon = \frac{p_z^2}{2m} + \hbar\omega_c \left(n + \frac{1}{2}\right) \le \epsilon_F. \qquad (9.8.11)$$

Going over from integration over p_z to that over ϵ with the above limit, we find

$$\int \frac{\Delta\Gamma}{(2\pi\hbar)^3} f(\epsilon) = \frac{1}{2} (2m)^{\frac{3}{2}} \frac{V g_S \omega_c}{(2\pi\hbar)^2} \int_0^{\epsilon_F} \sum_n \frac{f(\epsilon)}{\sqrt{\epsilon - \hbar\omega_c \left(n + \frac{1}{2}\right)}} d\epsilon. \qquad (9.8.12)$$

We must have restriction on the maximum allowed value of n so that the discriminant in the denominator must be positive. We now interchange the order of the integration and the summation and finally obtain

$$\int \frac{\Delta\Gamma}{(2\pi\hbar)^3} f(\epsilon) = \frac{1}{2} (2m)^{\frac{3}{2}} \frac{V g_S \omega_c}{(2\pi\hbar)^2} \sum_{n=0}^{\nu} \int_{(n+\frac{1}{2})\hbar\omega_c}^{\epsilon_F} \frac{f(\epsilon)}{\sqrt{\epsilon - \hbar\omega_c \left(n + \frac{1}{2}\right)}} d\epsilon, \qquad (9.8.13)$$

where $\nu = \left[\frac{\epsilon_F}{\hbar\omega_c} - \frac{1}{2}\right]$ is the integer part of the number $\frac{\epsilon_F}{\hbar\omega_c} - \frac{1}{2}$.

Thus energy of the system will be

$$E = \frac{1}{2} (2m)^{\frac{3}{2}} \frac{V g_S \omega_c}{(2\pi\hbar)^2} \sum_{n=0}^{\nu} \int_{(n+\frac{1}{2})\hbar\omega_c}^{\epsilon_F} \frac{\epsilon}{\sqrt{\epsilon - \hbar\omega_c \left(n + \frac{1}{2}\right)}} d\epsilon$$

$$= \frac{3N}{4} \frac{\hbar\omega_c}{\epsilon_F^{\frac{3}{2}}} \sum_{n=0}^{\nu} \int_{(n+\frac{1}{2})\hbar\omega_c}^{\epsilon_F} \frac{\epsilon}{\sqrt{\epsilon - \hbar\omega_c \left(n + \frac{1}{2}\right)}} d\epsilon \qquad (9.8.14)$$

The integration can be analytically evaluated by substituting $\epsilon = (n + \frac{1}{2})\hbar\omega_c \cosh^2 z$ giving us

$$E = \frac{3N}{4} \frac{\hbar\omega_c}{\epsilon_F^{\frac{3}{2}}} \sum_{n=0}^{\nu} \left[\epsilon_F \left(\epsilon_F - \hbar\omega_c \left(n + \frac{1}{2}\right)\right)^{\frac{1}{2}} - \frac{2}{3} \left(\epsilon_F - \hbar\omega_c \left(n + \frac{1}{2}\right)\right)^{\frac{3}{2}}\right]. \qquad (9.8.15)$$

We evaluate the summations by the result of Equation 14.12.4 of § 14 and keep terms up to B^2

$$E = \frac{3N}{4} \frac{\hbar\omega_c}{\epsilon_F^{\frac{3}{2}}} \times \left[\frac{1}{12} \hbar\omega_c \sqrt{\epsilon_F}\right] = \frac{N}{4} \frac{\mu_0^2 B^2}{\epsilon_F}, \qquad (9.8.16)$$

where $\mu_0 = \frac{q\hbar}{2mc}$ is the orbital magnetic moment of the particle.

Since at $T = 0$, $F = E - TS$, we obtain for magnetization defined as total magnetic moment per unit volume

$$\mathbf{M} = -\frac{1}{V}\frac{\partial F}{\partial \mathbf{B}} = -\frac{1}{V}\frac{\partial E}{\partial \mathbf{B}} = -\frac{1}{2}\frac{N}{V}\frac{\mu_0^2}{\epsilon_F}\mathbf{B} = \chi_L \mathbf{B}, \tag{9.8.17}$$

giving rise to the **Landau Susceptibility**

$$\chi_L = -\frac{1}{2}\frac{N}{V}\frac{\mu_0^2}{\epsilon_F}. \tag{9.8.18}$$

Since for electrons $\mu_0 = \mu_B$, the *Bohr Magneton*, we correspondingly have

$$\chi_L^{el} = -\frac{1}{2}\frac{N}{V}\frac{\mu_B^2}{\epsilon_F}. \tag{9.8.19}$$

The negative sign establishes that this is indeed a diamagnetic system. Anticipating the result of § 9.8.3 we find that Landau susceptibility is one-third in magnitude of the Pauli susceptibility.

9.8.3 *Pauli Paramagnetism*

In the absence of the external field at $T = 0$, $\frac{N}{2}$ particles have their magnetic moment $\vec{\mu_0}$ parallel to the $z-$axis with component $+\mu_0$, and the remaining $\frac{N}{2}$ particles have their magnetic moment anti-parallel to the $z-$axis with component $-\mu_0$. For simplicity we are assuming that the components can have *only two* values. The two types of particles have the common Fermi Energy

$$\epsilon_F = \frac{\hbar^2}{2m}\left(\frac{6\pi^2}{g_S}\right)^{\frac{2}{3}}\left(\frac{N}{2V}\right)^{\frac{2}{3}}. \tag{9.8.20}$$

Since every microstate can accommodate only one fermion, the density of state given by Equation 9.7.8 also denotes the number of total particles per unit energy interval with energy ϵ. Number of particles with either $+\mu_0$ or $-\mu_0$ moment component per unit energy interval at energy ϵ_F is thus

$$\left(\frac{dn_\pm(\epsilon)}{d\epsilon}\right)_{\epsilon_F} = \frac{3}{4}\frac{N}{\epsilon_F}. \tag{9.8.21}$$

When the system is placed in the uniform external field $\mathbf{B}$ along the $z-$axis, particles with their magnetic moment parallel to $\mathbf{B}$ will have an extra energy $-\mu_0 B$ and those with their magnetic moment anti-parallel to $\mathbf{B}$ will have extra energy $+\mu_0 B$. The resulting position is shown in Figure 9.8.1. Equilibrium will be established by the transfer of particles

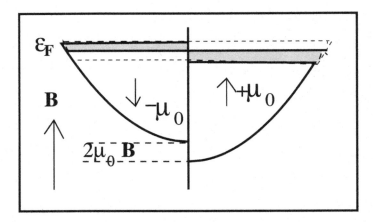

Fig. 9.8.1   Plot of the densities of state $\frac{dn(\epsilon)}{d\epsilon}$ for the two alignments of the intrinsic spin with respect to the uniform external magnetic field **B** at zero temperature.

of anti-parallel moments in the energy shell of width $\mu_0 B$ around ϵ_F to states with parallel moments, so that both the types of particles will again have the same Fermi Energy. For each particle the resulting change of magnetic moment is $2\mu_0$, and the total number of particles involved is $\left(\frac{dn_\pm(\epsilon)}{d\epsilon}\right)_{\epsilon_F} (\mu_0 B)$. From this total change of magnetic moment we get for magetization defined as magnetic moment per unit volume

$$\mathbf{M} = \frac{3}{2}\frac{N}{V}\frac{\mu_0^2}{\epsilon_F}\mathbf{B}. \qquad (9.8.22)$$

Thus we obtain the expression for **Pauli Susceptibility** for fermions in general and for electrons in particular with $\mu_0 = \mu_B$, the **Bohr Magneton**,

$$\chi_P = \frac{3}{2}\frac{N}{V}\frac{\mu_0^2}{\epsilon_F}, \qquad (9.8.23)$$

$$\chi_P^{\text{el}} = \frac{3}{2}\frac{N}{V}\frac{\mu_B^2}{\epsilon_F}. \qquad (9.8.24)$$

The positive sign of χ_P proves that it is a case of paramagnetism.

When the inverse susceptibility $\frac{1}{\chi}$ of the fermion system is plotted against temperature T as shown in Figure 9.8.2 the Curie's law behaviour $\chi \propto \frac{1}{T}$ is obtained at high temperature limit while the asymptotic value at $T = 0$ will give the Pauli value χ_P.

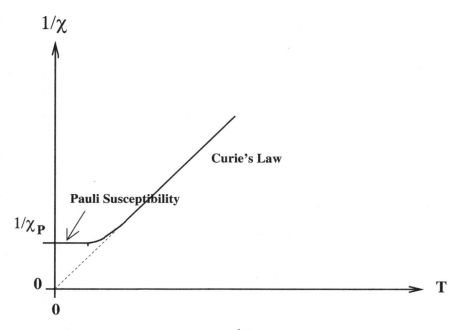

$1/\chi$

Curie's Law

Pauli Susceptibility

$1/\chi_P$

0

0

T

Fig. 9.8.2 Plot of the inverse susceptibility $\frac{1}{\chi}$ as a function of temperature. The high temperature linear part shows Curie law behaviour and the zero-temperature value gives Pauli susceptibility.

9.9 Interacting Fermi System: Fermi Liquid Theory

The Landau Fermi Liquid Theory is the most powerful tool for studying interacting Fermi systems. A similar analysis of interacting Bose systems in terms of the phonon and roton excitations in liquid ^4He has been described in § 9.6.2. In the same spirit the low-lying excited states of interacting fermi systems are described by the presence of excitations called **quasiparticles**. If in the ground state of the system the number of quasi-particle with momentum $\mathbf{p}$ and spin σ changes by an amount $\delta n_{\mathbf{p},\sigma}$ the change in energy of the system, in the absence of interaction between quasiparticles, is written as

$$\delta E = \sum_{\mathbf{p},\sigma} \epsilon_{\mathbf{p},\sigma} \delta n_{\mathbf{p},\sigma}, \tag{9.9.1}$$

so that $\epsilon(\mathbf{p}, \sigma)$ is the functional derivative

$$\epsilon_{\mathbf{p},\sigma} = \frac{\delta E}{\delta n_{\mathbf{p},\sigma}}, \tag{9.9.2}$$

defined in § 14.3. The thermal equilibrium value of the occupancy of the quasiparticles are obtained from separately equating the coefficients of different $\delta n(\mathbf{p}, \sigma)$ from both sides in

$$\delta E = T\delta S + \mu\delta n, \tag{9.9.3}$$

$$\sum_{\mathbf{p},\sigma} \epsilon_{\mathbf{p},\sigma}\delta n_{\mathbf{p},\sigma} = -k_B T \sum_{\mathbf{p},\sigma} \log\left(\frac{n_{\mathbf{p},\sigma}}{1-n_{\mathbf{p},\sigma}}\right)\delta n_{\mathbf{p},\sigma} + \mu \sum_{\mathbf{p},\sigma} \delta n_{\mathbf{p},\sigma},$$

$$\epsilon_{\mathbf{p},\sigma} = \mu + k_B T \log\left(\frac{1}{n_{\mathbf{p},\sigma}} - 1\right),$$

$$n_{\mathbf{p},\sigma} = \frac{1}{\exp\left(\frac{\epsilon_{\mathbf{p},\sigma}-\mu}{k_B T}\right) + 1}. \tag{9.9.4}$$

This shows that the quasiparticles also are fermions.

We now assume that there are collisions between the quasi-particles and the energy functional becomes

$$\delta E = \sum_{\mathbf{p},\sigma} \epsilon_{\mathbf{p},\sigma}\delta n_{\mathbf{p},\sigma} + \frac{1}{2} \sum_{\substack{\mathbf{p},\mathbf{p}' \\ \sigma,\sigma'}} f_{\mathbf{p},\sigma;\mathbf{p}',\sigma'}\delta n_{\mathbf{p},\sigma}\delta n_{\mathbf{p}',\sigma'}, \tag{9.9.5}$$

$$f_{\mathbf{p},\sigma;\mathbf{p}',\sigma'} = \frac{\delta^2 E}{\delta n_{\mathbf{p},\sigma}\delta n_{\mathbf{p}',\sigma'}}. \tag{9.9.6}$$

The energy for the 'dressed' quasi-particles now becomes

$$\tilde{\epsilon}_{\mathbf{p},\sigma} = \frac{\delta E}{\delta n_{\mathbf{p},\sigma}} = \epsilon_{\mathbf{p},\sigma} + \sum_{\mathbf{p}',\sigma'} f_{\mathbf{p},\sigma;\mathbf{p}',\sigma'}\delta n_{\mathbf{p}',\sigma'}, \tag{9.9.7}$$

and they move about in the medium with an effective mass m^* which is a result of the interaction between the quasi-particles. We now proceed to evaluation of this effective mass m^* by calculating mass flux of the quasi-particles.

$$\sum_{\mathbf{p},\sigma} \frac{\mathbf{p}}{m} n_{\mathbf{p},\sigma} = \sum_{\mathbf{p},\sigma} \left(\frac{\partial \tilde{\epsilon}_{\mathbf{p},\sigma}}{\partial \mathbf{p}} \right) n_{\mathbf{p},\sigma},$$

$$\sum_{\mathbf{p},\sigma} \frac{\mathbf{p}}{m} \delta n_{\mathbf{p},\sigma} = \sum_{\mathbf{p},\sigma} \left(\left(\frac{\partial \delta \tilde{\epsilon}_{\mathbf{p},\sigma}}{\partial \mathbf{p}} \right) n_{\mathbf{p},\sigma} + \left(\frac{\partial \tilde{\epsilon}_{\mathbf{p},\sigma}}{\partial \mathbf{p}} \right) \delta n_{\mathbf{p},\sigma} \right)$$

$$= \sum_{\mathbf{p},\sigma} n_{\mathbf{p},\sigma} \left(\sum_{\mathbf{p}',\sigma'} \frac{\partial f_{\mathbf{p},\sigma;\mathbf{p}',\sigma'}}{\partial \mathbf{p}} \delta n_{\mathbf{p}'.\sigma'} \right) + \sum_{\mathbf{p},\sigma} \frac{\partial \tilde{\epsilon}_{\mathbf{p},\sigma}}{\partial \mathbf{p}} \delta n_{\mathbf{p},\sigma}$$

$$= -\sum_{\mathbf{p},\sigma} \frac{\partial n_{\mathbf{p},\sigma}}{\partial \mathbf{p}} \sum_{\mathbf{p}',\sigma'} f_{\mathbf{p},\sigma;\mathbf{p}',\sigma'} \delta n_{\mathbf{p}'.\sigma'} + \sum_{\mathbf{p},\sigma} \frac{\partial \tilde{\epsilon}_{\mathbf{p},\sigma}}{\partial \mathbf{p}} \delta n_{\mathbf{p},\sigma}$$

$$= -\sum_{\mathbf{p}',\sigma'} \frac{\partial n_{\mathbf{p}',\sigma'}}{\partial \mathbf{p}'} \sum_{\mathbf{p},\sigma} f_{\mathbf{p}',\sigma';\mathbf{p},\sigma} \delta n_{\mathbf{p}.\sigma} + \sum_{\mathbf{p},\sigma} \frac{\partial \tilde{\epsilon}_{\mathbf{p},\sigma}}{\partial \mathbf{p}} \delta n_{\mathbf{p},\sigma}$$

$$= \sum_{\mathbf{p},\sigma} \delta n_{\mathbf{p},\sigma} \left[\frac{\partial \tilde{\epsilon}_{\mathbf{p},\sigma}}{\partial \mathbf{p}} - \sum_{\mathbf{p}',\sigma'} f_{\mathbf{p}',\sigma';\mathbf{p},\sigma} \frac{\partial n_{\mathbf{p}',\sigma'}}{\partial \mathbf{p}'} \right]$$

$$= \sum_{\mathbf{p},\sigma} \delta n_{\mathbf{p},\sigma} \left[\frac{\partial \tilde{\epsilon}_{\mathbf{p},\sigma}}{\partial \mathbf{p}} - \sum_{\mathbf{p}',\sigma'} f_{\mathbf{p}',\sigma';\mathbf{p},\sigma} \frac{\partial n_{\mathbf{p}',\sigma'}}{\partial \tilde{\epsilon}_{\mathbf{p}',\sigma'}} \frac{\partial \tilde{\epsilon}_{\mathbf{p}',\sigma'}}{\partial \mathbf{p}'} \right]. \tag{9.9.8}$$

Equating coefficients of each $\delta n_{\mathbf{p},\sigma}$ from both sides of the equation and using $\frac{\partial \tilde{\epsilon}_{\mathbf{p},\sigma}}{\partial \mathbf{p}} = \frac{\mathbf{p}}{m^*}$, we get

$$\frac{\mathbf{p}}{m} = \frac{\mathbf{p}}{m^*} - \sum_{\mathbf{p}',\sigma'} f_{\mathbf{p}'.\sigma';\mathbf{p},\sigma} \frac{\partial n_{\mathbf{p}',\sigma'}}{\partial \tilde{\epsilon}_{\mathbf{p}',\sigma'}} \frac{\mathbf{p}'}{m^*}$$

$$\frac{1}{m} = \frac{1}{m^*} \left[1 - \sum_{\mathbf{p}',\sigma'} f_{\mathbf{p}'.\sigma';\mathbf{p},\sigma} \frac{\partial n_{\mathbf{p}',\sigma'}}{\partial \tilde{\epsilon}_{\mathbf{p}',\sigma'}} \frac{p'}{p} \left(\hat{\mathbf{p}} \cdot \hat{\mathbf{p}}' \right) \right]. \tag{9.9.9}$$

The collision parameters are expanded in terms of Legendre polynomials $P_l(\cos \theta)$, where $\theta =$ the angle between $\hat{\mathbf{p}}$ and $\hat{\mathbf{p}}'$,

$$f_{\mathbf{p}\sigma,\mathbf{p}'\sigma'} = \sum_{l=0}^{\infty} \left(f_l^s + \vec{\sigma} \cdot \vec{\sigma}' f_l^a \right) P_l(\cos \theta) = \sum_{l=0}^{\infty} \left(f_l^s + \vec{\sigma} \cdot \vec{\sigma}' f_l^a \right) P_l \left(\hat{\mathbf{p}} \cdot \hat{\mathbf{p}}' \right), \tag{9.9.10}$$

where f_l^s are the spin-independent spherical parts and f_l^a are the spin-dependent axial parts.

The summation over $\mathbf{p}'$ can be transformed to integration over energy in the form

$$\frac{1}{m} = \frac{1}{m^*} \left[1 - \sum_{l=0}^{\infty} \sum_{\sigma'} \int_0^{\epsilon_F} \frac{\pi V (2m^*)^{\frac{3}{2}} \sqrt{\tilde{\epsilon}'}}{(2\pi\hbar)^3} \sqrt{\frac{\tilde{\epsilon}'}{\tilde{\epsilon}}} \left[-\delta \left(\tilde{\epsilon}' - \epsilon_F \right) \right] d\tilde{\epsilon}'$$

$$\times \int_{\theta=0}^{\theta=\pi} \left(f_l^s + \left(\vec{\sigma} \cdot \vec{\sigma}' \right) f_l^a \right) \cos\theta P_l \left(\cos\theta \right) \sin\theta d\theta \right].$$

$$(9.9.11)$$

Here we have used the $T = 0 \ K$ value of $\frac{dn(\tilde{\epsilon})}{d\tilde{\epsilon}} = -\delta(\tilde{\epsilon} - \epsilon_F)$ and the expression $\tilde{\epsilon}(\mathbf{p}) = \frac{p^2}{2m^*}$ for the energy of the dressed quasi-particles.

Performing the energy integral and using the expression for the Fermi Energy $\epsilon_F = \frac{\hbar^2}{2m^*} \left(\frac{6\pi^2}{g_s} \right)^{\frac{2}{3}} \left(\frac{N}{V} \right)^{\frac{2}{3}}$, we obtain

$$\frac{1}{m} = \frac{1}{m^*} \left[1 + \frac{3N}{4\epsilon_F} \sum_{l=0}^{\infty} \int_{\theta=0}^{\theta=\pi} \sum_{\sigma'} \left(f_l^s + \left(\vec{\sigma} \cdot \vec{\sigma}' \right) f_l^a \right) \cos\theta P_l \left(\cos\theta \right) \sin\theta d\theta \right]$$

$$= \frac{1}{m^*} \left[1 + \frac{1}{4} \sum_{l=0}^{\infty} \int_{\theta=0}^{\theta=\pi} \sum_{\sigma'} \left(F_l^s + \left(\vec{\sigma} \cdot \vec{\sigma}' \right) F_l^a \right) \cos\theta P_l \left(\cos\theta \right) \sin\theta d\theta \right].$$

$$(9.9.12)$$

Here we define the **Landau Parameters** as

$$F_l^{s,a} = \frac{3N}{\epsilon_F} f_l^{s,a}. \qquad (9.9.13)$$

Only the spherical parts F_l^s contributes to thermal or other spin-independent properties like Heat Capacity while the axial part F_l^a contributes to spin dependent properties like magnetic susceptibility. Because of the orthogonality property of the Legendre polynomial only $l = 1$ term contributes to the effective mass through F_l^s, and after the integration over θ we arrive at

$$\frac{1}{m} = \frac{1}{m^*} \left[1 + \frac{1}{3} F_1^s \right]. \qquad (9.9.14)$$

For repulsive interaction F_1^s is positive and hence the effective mass $m^* > m$, the bare mass. Since Heat capacity is directly proportional to mass, we see that $C_{\text{FL}} > C_{\text{free}}$. Fermi Liquid theory has been successfully aplied to explain the properties of Liquid ^3He and Heavy Fermion systems like $CeAl_3$ and $CeCu_2Si_2$.

9.10 Relativistic Degenerate Fermi Gas

For a relativistic degenerate fermi gas of rest mass m the energy spectrum is

$$\epsilon(p) = \sqrt{p^2c^2 + (mc^2)^2}. \tag{9.10.1}$$

Hence the minimum value of energy is mc^2. So at $T = 0$ the occupancy

$$\overline{n}(\epsilon) = \lim_{T \to 0} \frac{1}{\exp\left(\frac{\epsilon - \mu}{k_B T}\right) + 1} \tag{9.10.2}$$

is as shown in Figure 9.10.1. The results of § 9.7 can be used after putting

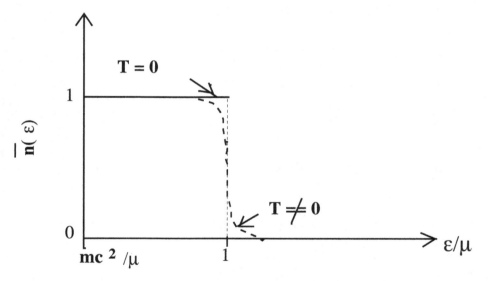

Fig. 9.10.1   Occupancy $\overline{n}(\epsilon)$ for relativistic degenerate fermi system of rest mass m at zero degree Kelvin and at non-zero Temperature.

$\epsilon_0 = mc^2$. In particular, the expressions for particle density and energy density in this case will become

$$\frac{N}{V} = \frac{4\pi g}{(2\pi\hbar c)^3} \int_{mc^2}^{\infty} \frac{\epsilon\sqrt{\epsilon^2 - (mc^2)^2}}{\exp\left(\frac{\epsilon - \mu}{k_B T}\right) + 1} \, d\epsilon$$

$$= \frac{4\pi g}{(2\pi\hbar c)^3} \left[\int_{mc^2}^{\mu} \epsilon\sqrt{\epsilon^2 - (mc^2)^2} d\epsilon \right.$$

$$+ 2\sum_{n=0}^{\infty} (k_B T)^{2n+2} \left(\left(\frac{d}{d\mu}\right)^{2n+1} \mu\sqrt{\mu^2 - (mc^2)^2} \right)$$

$$\left. \times \left(1 - \frac{1}{2^{2n+1}} \right) \zeta(2n+2) \right], \quad (9.10.3)$$

$$\frac{E}{V} = \frac{4\pi g}{(2\pi\hbar c)^3} \int_{mc^2}^{\infty} \frac{\epsilon^2\sqrt{\epsilon^2 - (mc^2)^2}}{\exp\left(\frac{\epsilon - \mu}{k_B T}\right) + 1} \, d\epsilon$$

$$= \frac{4\pi g}{(2\pi\hbar c)^3} \left[\int_{mc^2}^{\mu} \epsilon^2\sqrt{\epsilon^2 - (mc^2)^2} d\epsilon \right.$$

$$+ 2\sum_{n=0}^{\infty} (k_B T)^{2n+2} \left(\left(\frac{d}{d\mu}\right)^{2n+1} \mu^2\sqrt{\mu^2 - (mc^2)^2} \right)$$

$$\left. \times \left(1 - \frac{1}{2^{2n+1}} \right) \zeta(2n+2) \right]. \quad (9.10.4)$$

The integrals appearing in Equations 9.10.3 and 9.10.4 can be evaluated by substituting $\epsilon = mc^2 \cosh t$,

$$\int_{mc^2}^{\mu} \epsilon\sqrt{\epsilon^2 - (mc^2)^2} d\epsilon = \frac{1}{3}\left(\mu^2 - (mc^2)^2\right)^{\frac{3}{2}}, \quad (9.10.5)$$

$$\int_{mc^2}^{\mu} \epsilon^2\sqrt{\epsilon^2 - (mc^2)^2} d\epsilon = (mc^2)^4 \left[\frac{1}{32}\sinh(4t_m) - \frac{1}{8}t_m \right], \quad (9.10.6)$$

where

$$\cosh t_m = \frac{\mu}{mc^2}. \quad (9.10.7)$$

The zero temperature value for Equation 9.10.3 will give us the Fermi Energy

$$\epsilon_F = mc^2\sqrt{1 + \left(\frac{\hbar}{mc}\right)^2 \left(\frac{6\pi^2}{g}\right)^{\frac{2}{3}} \left(\frac{N}{V}\right)^{\frac{2}{3}}}. \quad (9.10.8)$$

We have expressed energy in terms of the rest-mass energy mc^2 and length in terms of the **deBroglie wavelength** $\lambda_D = \frac{\hbar}{mc}$ of the particle. The temperature dependence of the chemical potential will become

$$\mu\left(\frac{N}{V}, T\right) = \epsilon_F \left[1 - \frac{\pi^2}{6}\left(\frac{k_B T}{\epsilon_F}\right)^2 \left(2 + \left(\frac{mc}{\hbar}\right)^2 \left(\frac{g}{6\pi^2}\right)^{\frac{2}{3}} \left(\frac{V}{N}\right)^{\frac{2}{3}}\right) + \cdots\right].$$
(9.10.9)

The energy density at $T = 0\ K$ is given by

$$\frac{E_0}{V} = 3\left(\frac{g}{6\pi^2}\right)\left(\frac{mc}{\hbar}\right)^3 mc^2 \left[\frac{1}{32}\sinh\left(4t_m\right) - \frac{1}{8}t_m\right].$$
(9.10.10)

Using the relativistic energy spectrum and Equation 9.2.3 we obtain expression for pressure

$$
\begin{aligned}
P &= -\frac{\Omega}{V} = k_B T \frac{4\pi g}{(2\pi\hbar)^3}\int_0^\infty p^2 \log\left[1 + \exp\left(\frac{\mu - \epsilon(p)}{k_B T}\right)\right] dp \\
&= k_B T \frac{4\pi g}{(2\pi\hbar c)^3}\int_{mc^2}^\infty \log\left[1 + \exp\left(\frac{\mu - \epsilon}{k_B T}\right)\right]\epsilon\sqrt{\epsilon^2 - (mc^2)^2}\, d\epsilon \\
&= \frac{4\pi g}{(2\pi\hbar c)^3}\int_{mc^2}^\infty d\epsilon\, \frac{\exp\left(\frac{\mu - \epsilon}{k_B T}\right)}{1 + \exp\left(\frac{\mu - \epsilon}{k_B T}\right)}\left(\int \epsilon\sqrt{\epsilon^2 - (mc^2)^2}\, d\epsilon\right) \\
&= \frac{4\pi g}{(2\pi\hbar c)^3}\int_{mc^2}^\infty \frac{\frac{1}{3}\left(\epsilon^2 - (mc^2)^2\right)^{\frac{3}{2}}}{\exp\left(\frac{\epsilon - \mu}{k_B T}\right) + 1}\, d\epsilon \\
&= \frac{1}{3}\frac{4\pi g}{(2\pi\hbar c)^3}\int_{mc^2}^\infty \frac{\epsilon^2\sqrt{\epsilon^2 - (mc^2)^2}}{\exp\left(\frac{\epsilon - \mu}{k_B T}\right) + 1}\, d\epsilon \\
&\qquad - \frac{1}{3}\frac{4\pi g}{(2\pi\hbar c)^3}(mc^2)^2\int_{mc^2}^\infty \frac{\sqrt{\epsilon^2 - (mc^2)^2}}{\exp\left(\frac{\epsilon - \mu}{k_B T}\right) + 1}\, d\epsilon \\
&= \frac{1}{3}\frac{E}{V} \\
&\qquad - \left(\frac{g}{6\pi^2}\right)\left(\frac{mc}{\hbar}\right)^3 (mc^2)\left[\int_{mc^2}^\infty \sqrt{\epsilon^2 - (mc^2)^2}\, d\epsilon\right. \\
&\qquad + 2\sum_{n=0}^\infty (k_B T)^{2n+2}\left(\left(\frac{d}{d\mu}\right)^{2n+1}\sqrt{\mu^2 - (mc^2)^2}\right) \\
&\qquad \left.\times \left(1 - \frac{1}{2^{2n+1}}\right)\zeta(2n+2)\right].
\end{aligned}
$$
(9.10.11)

After evaluating the integral we get for zero-temperature pressure for a relativistic free fermion system

$$P_0 = \left(\frac{g}{6\pi^2}\right)\left(\frac{mc}{\hbar}\right)^3 (mc^2)\left[\frac{1}{32}\sinh\left(4t_m\right) - \frac{1}{4}\sinh\left(2t_m\right) + \frac{3}{8}t_m\right],$$
(9.10.12)

with

$$\cosh\left(t_m\right) = \frac{\epsilon_F}{mc^2}.$$
(9.10.13)

9.11   Problems

Problem 9.1.   Prove Equation 9.2.4.

Problem 9.2.   Obtain the expression for density of state for a system of non-relativistic particles of mass m moving in a 2-dimensional thin film of area A.

Problem 9.3.   Obtain the expression for density of state for a system of relativistic particles of rest mass m moving in a 2-dimensional thin film of area A.

Problem 9.4.   Obtain the expression for density of state for a system of non-relativistic particles of mass m moving in a 1-dimensional thin film of length L.

Problem 9.5.   Obtain the expression for density of state for a system of relativistic particles of rest mass m moving in a 2-dimensional thin film of area A.

Problem 9.6.   Prove Equation 9.2.18.

Problem 9.7.   Prove Equations 9.2.19, 9.2.20, 9.2.21 and 9.2.22.

Problem 9.8.   Show that for both the BE and the FD statistics a non-relativistic quantum system of particles with mass m moving in a 2-dimensional thin film of area A will have the Surface Tension $\mathcal{S} = \frac{E}{A}$, where E is the energy of the system.

Problem 9.9.   Show that for both the BE and the FD statistics a non-relativistic quantum system of particles with mass m moving in a 1-dimensional linear ploymer of length L will have the Tension $\mathcal{F} = 2\frac{E}{L}$, where E is the energy of the system.

Problem 9.10. Calculate the second Virial coefficient $B(T)$ for a non-relativistic quantum system of particles with mass m moving in a 2-dimensional thin film of area A.

Problem 9.11. Calculate the second Virial coefficient $B(T)$ for a non-relativistic quantum system of particles with mass m moving in a 1-dimensional linear polymer of area L.

Problem 9.12. Prove Equation 9.2.24.

Problem 9.13. Prove Equation 9.3.11.

Problem 9.14. Prove Equations 9.4.7 and 9.4.10.

Problem 9.15. Prove Equations 9.5.1, 9.5.2 and 9.5.4.

Problem 9.16. Prove Equation 9.6.1.

Problem 9.17. Prove Equations 9.6.10, 9.6.11 and 9.6.12.

Problem 9.18. Prove Equations 9.6.13, 9.6.14 and 9.6.15.

Problem 9.19. Prove Equation 9.6.20.

Problem 9.20. Calculate the Fermi Energy ϵ_F in terms of the number density $\frac{N}{A}$ for a non-relativistic quantum system of particles with mass m moving in a 2-dimensional thin film of area A.

Problem 9.21. Calculate the Fermi Energy ϵ_F in terms of the number density $\frac{N}{L}$ for a non-relativistic quantum system of particles with mass m moving in a 1-dimensional linear polymer of length L.

Problem 9.22. Prove Equations 9.7.11 and 9.7.12.

Problem 9.23. Prove Equations 9.8.16 and 9.8.18.

Problem 9.24. Show that the magnetic susceptibility in the Fermi Liquid theory is given by $\chi_{\text{FL}} = \chi_{\text{P}}^{\text{el}} \, \frac{m^*/m}{1+F_0^a}$.

Problem 9.25. Prove Equations 9.10.3 and 9.10.4.

Problem 9.26. Prove Equations 9.10.9 and 9.10.10.

Problem 9.27. Prove Equations 9.10.11 and 9.10.12.

Chapter 10

Bose-Einstein Condensate

10.1 Introduction

In § 9.5 and § 9.6 we have seen how a Bose gas of non-interacting particles undergoes Bose-Einstein (BE) condensation and how the superfluid phase of liquid Helium was supposed to be such a condensed state. This was in spite of the facts that a liquid is not a non-interacting system and neutron scattering experiments definitely show that near 0 K only 10% of helium particles occupy the zero-momentum state. Moreover near the transition temperature specific heat of a non-interacting system does *not* have a singularity, while liquid helium has the famous λ-point singularity. Discovery in 1995 of **Bose-Einstein Condensation** (BEC) in trapped atomic cloud of ^{87}Rb was the first experimental confirmation of this interesting physical phenomenon. Neutral atoms have the same number of protons and electrons; so the condition that the atom will obey Bose statistics is given by the evenness of the number of neutrons in the isotope.

In a gas the average interparticle distance is $\propto n^{-\frac{1}{3}}$ where n is the number density. At high temperature the **thermal de Broglie wavelength**,

$$\lambda_{\mathrm{dB}} = h\sqrt{\frac{2\pi}{mk_BT}} \ll n^{-\frac{1}{3}} \qquad (10.1.1)$$

and the system behaves as a classical gas. Bose condensation sets in when $\lambda_{\mathrm{dB}} \sim n^{-\frac{1}{3}}$. With the typical experimental set-up with $n \sim 10^{12}$ cm^{-3} the condensation temperature $T_c \sim 100\ nK$.

We had to wait so long for creating true BE condensation because new methods for creating ultracold temperature and trapping the atoms in a small region of space without requiring a cryostat by using laser was developed only very recently.

10.2   Trapping of Atoms

The trapping of the alkali atom vapor in a small region of space is done by quadrupolar magnetic trap with a spatially varying magnetic field. The effect of the magnetic field on the atom is through the *Zeeman term* $\mathcal{H}_Z = CJ_z + DI_z$ and the *hyperfine interaction* $\mathcal{H}_{hf} = A\mathbf{I} \cdot \mathbf{J}$. Here $\mathbf{I}$ and $\mathbf{J}$ are the intrinsic spins of the nucleus and the electrons respectively. For ^{87}Rb with 50 neutrons, 37 protons and 37 electrons in electronic configuration $[Ar]3d^{10}4s^24p^65s^1$ $I = \frac{3}{2}$ and $J = \frac{1}{2}$ there is no complication due to orbital angular momentum of the electrons. The $\mathcal{H}_{hf}$ splits the ground-state multiplet in a 3-fold degenerate ground state $|e_g\rangle$ with energy $E_g = -\frac{5A}{4}$ and total angular momentum $F = 1$, and a 5-fold degenerate excited state $|e_{exc}\rangle$ with energy $E_{exc} = +\frac{3A}{4}$ and $F = 2$. The energy separation is $\Delta = 2A = h\Delta\nu$. For ^{87}Rb the measured value is $\Delta\nu = 6834.683$MHz. Zeeman splitting of the energies with increasing magnetic field is shown in Figure 10.2.1. Trapping of atomic clouds in spatially varying magnetic field

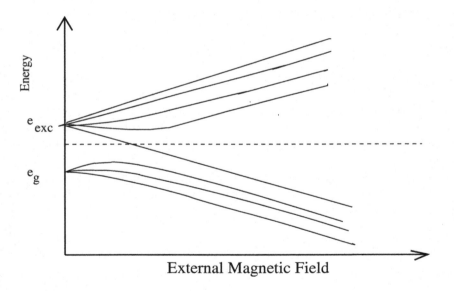

Fig. 10.2.1   Schematic plot of energies of hyperfine levels in external magnetic field.

makes use of this variation of energies. In the quadrupolar configuration (Figure 10.2.2) a field gradient $-2B'$ along the z-direction is generated by

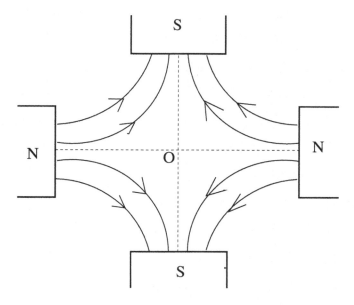

Fig. 10.2.2   Schematic drawing of a quadrupolar trapping configuration.

$$\mathbf{B} = \begin{pmatrix} B'x \\ B'y \\ -2B'z \end{pmatrix} \qquad (10.2.1)$$

with the minimum at the origin O. For slow *adiabatic* motion of the atoms the atomic states are not changed. Atoms in the state whose energy increases with increasing magnetic field will seek the region where magnetic field is minimum. These are called '**low-field seekers**'. Those in states whose energy decreases with increasing magnetic field will come to the region where magnetic field is maximum. These are called '**high-field seekers**'. Thus in a quadrupolar trap atoms in 'high-field seeking' state will be expelled from the trap.

Moreover, as a result of collision atoms in 'low-field seeking' state may make a transition to the 'high-field seeking' state and thus leaves the trap depleting the atomic concentration. However, the state where the component of both the electronic and the nuclear spin has the maximum value is particularly stable. For ^{87}Rb the state $|m_J = \frac{1}{2}, m_I = \frac{3}{2}\rangle$ is such a state.

Near the node of the magnetic field, the separation of energy between different magnetic states vanishes and slight perturbation induces transition between the substates. Thus there is a 'hole' in the trap near such a node. One of the way of 'plugging' this 'hole' is by using a 'time-averaged orbiting potential' or TOP trap. To plug the trap a spatially uniform, rotating magnetic field of constant strength is superimposed on the quadrupolar field, so that the instantaneous field is

$$\mathbf{B} = \begin{pmatrix} B'x + B_0 \cos \omega t \\ B'y + B_0 \sin \omega t \\ -2B'z \end{pmatrix}. \tag{10.2.2}$$

The instantaneous magnitude of the total field is

$$B(t) = \left[(B_0 \cos \omega t + B'x)^2 + (B_0 \sin \omega t + B'y)^2 + 4B'^2 z^2 \right]^{\frac{1}{2}}$$
$$\approx B_0 + B' (x \cos \omega t + y \sin \omega t)$$
$$+ \frac{B'^2}{2B_0} \left(x^2 + y^2 + 4z^2 - (x \cos \omega t + y \sin \omega t)^2 \right). \tag{10.2.3}$$

The time averaged magnitude $\overline{B}$ of the field near the node

$$\overline{B} = \frac{\omega}{2\pi} \int_0^{\frac{2\pi}{\omega}} B(t) dt \approx B_0 + \frac{B'^2}{4B_0} \left(x^2 + y^2 + 4z^2 \right) \tag{10.2.4}$$

is thus never zero and the 'hole' is effectively 'plugged'. The frequency ω should not be comparable to frquencies of transitions between magnetic states lest it induces transitions between these states leading to loss of atoms from the trap.

10.3   Cooling of Atoms

Cooling the atomic clouds to temperature $\sim 100 \ nK$ is done in two steps. In the first step atoms are cooled to $\sim 10 \ \mu K$ by laser cooling. The principle of this cooling method can be explained by considering the atom being impinged on by two oppositely moving laser beams as shown in Figure 10.3.1. For each beam rate of absorption of photon of frequency ω is given by

$$\frac{dn_{ph}}{dt} = KL(\omega) = \frac{1}{\tau}, \tag{10.3.1}$$

where

$$L(\omega) = \frac{\frac{\Gamma_e}{2\pi}}{\left(\omega - \omega_{eg} \right)^2 + \left(\frac{\Gamma_e}{2} \right)^2} \tag{10.3.2}$$

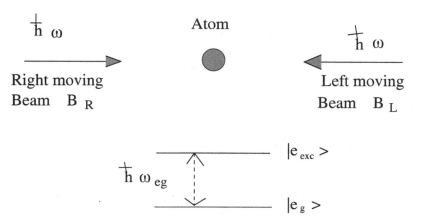

Fig. 10.3.1 Sketch of the method of laser cooling by two oppositely moving laser beams impinging on the atom of excited and ground states $|e_{exc}>$ and $|e_g>$ respectively.

is the normalized Lorentzian lineshape (vide § (14.7.2.4) of the Mathematical Appendix) and

τ = Lifetime of atom in ground state $|e_g\rangle$ in the presence of photon beam.

If the atom is at rest then the change in its total linear momentum is zero since it absorbs photons from both directions. If, however, it is moving to the right with velocity v_z, then in the rest frame of the atom the frequency of the right-moving beam will be Döppler shifted to $\omega_R = \omega \left(1 - \frac{v_z}{c}\right)$ and that of the left-moving beam will be $\omega_L = \omega \left(1 + \frac{v_z}{c}\right)$ and the absorption rate of the two types of photons will be

$$\frac{dn_R}{dt} = KL\left[\omega\left(1 - \frac{v_z}{c}\right)\right], \qquad (10.3.3)$$

$$\frac{dn_L}{dt} = KL\left[\omega\left(1 + \frac{v_z}{c}\right)\right]. \qquad (10.3.4)$$

Absorption of photon will be accompanied by absorption of linear momentum by the atom and the equation of motion of the atom will be

$$\frac{dp_z}{dt} = -\alpha v_z \qquad (10.3.5)$$

where

$$\alpha = \frac{\hbar\omega}{c}\frac{1}{v_z}K\left(L\left[\omega\left(1 + \frac{v_z}{c}\right)\right] - L\left[\omega\left(1 - \frac{v_z}{c}\right)\right]\right)$$

$$\approx \frac{2\hbar\omega^2}{c^2}K\frac{dL(\omega)}{d\omega}. \qquad (10.3.6)$$

This frictional force is quite large, since for the very narrow lines of lasers $\frac{dL(\omega)}{d\omega}$ is very large, and is responsible for creating what is known as **optical molasses**. The characteristic *braking time* τ_{fric} will satisfy

$$\frac{1}{\tau_{\text{fric}}} = -\frac{1}{p_z}\frac{dp_z}{dt} = \frac{\alpha}{m}. \tag{10.3.7}$$

Due to this frictional force the rate of change of the mean kinetic energy of the atom is

$$\left[\frac{d}{dt}\overline{p_z^2}\right]_{\text{fric}} = 2p_z\overline{\frac{dp_z}{dt}} = -\frac{2\alpha}{m}\overline{p_z^2}. \tag{10.3.8}$$

Since each absorbed photon imparts mpmentum $\frac{\hbar\omega}{c}$ to the atom and the rate of absorption of total number of photon from both the directions is in magnitude $\frac{dn_{\text{tot}}}{dt} = 2KL(\omega)$, we get after equating these two terms and using the law of equipartition of energy for 1-dimensional motion

$$k_B T = \frac{\hbar L(\omega)}{(dL(\omega)/d\omega)}, \tag{10.3.9}$$

where T is the absolute temperature. Lowest temperature is attained when $\frac{dT}{d\omega} = 0$. For the Lorentzian lineshape of the laser line this occurs when the frequency of the laser

$$\omega = \omega_{eg} - \frac{\Gamma_e}{2}. \tag{10.3.10}$$

is less than the resonance frequency of the atom. This condition is called **red detuning**. The lowest temperature attainable at the laser frequency satisfies this condition, is

$$T_{\min} = \frac{\hbar\Gamma_e}{2k_B}. \tag{10.3.11}$$

In the opposite case of **blue detuning** when laser frequency $\omega = \omega_{eg} + \frac{\Gamma_e}{2}$ the atomic vapor will be heated up instead of being cooled.

Bose condensation temperature $\sim 100\ nK$ is attained by further cooling of the atomic vapor by **evaporative cooling**. This is done by making a *hole* in the *trap* by applying an rf-field that can transform the atoms in the ground state to the excited state by flipping the spin states. Thus **low-field seekers** are transformed to **high field seekers**. This expels the high energy particles from the system and the system is cooled.

For a system of N number of atoms each of average rnergy ϵ the total energy of the system is $E = N\epsilon$. If a total of $dN < 0$ number of atoms each with an average energy $(1 + \mu)\epsilon$, where $\mu > 0$ is independent of N, is evaporated then after the evaporation

$$\text{Total number of atoms} = N + dN, \tag{10.3.12}$$

$$\text{Total Energy} = E + (1 + \mu)\epsilon dN, \tag{10.3.13}$$

$$\text{Average Energy} = \epsilon + d\epsilon = \frac{E + (1 + \mu)\,\epsilon dN}{N + dN}, \tag{10.3.14}$$

so that

$$\frac{d\epsilon}{\epsilon} = \frac{dN}{N}. \tag{10.3.15}$$

We thus get a power-law of decrease of average energy and hence of temperature according to

$$\frac{\epsilon}{\epsilon_0} = \left(\frac{N}{N_0}\right)^{\mu}, \tag{10.3.16}$$

where N_0 and ϵ_0 are the initial values when the evaporation process starts.

10.4   Problems

Problem 10.1.   Prove Equation 10.1.1.

Problem 10.2.   Prove Equations 10.3.5 and 10.3.6.

Problem 10.3.   Prove Equation 10.3.11.

Chapter 11

Statistical Astrophysics

11.1 Introduction

The objects studied in astrophysics are, among others, **stars, interstellar gas, nebulae, pulsars, quasars, neutron stars, black holes**. Some of these bodies are plotted in the *density-temperature phase diagram*, Figure 11.1.1. Neutron stars, Pulsars and White Dwarfs lie on the upper region

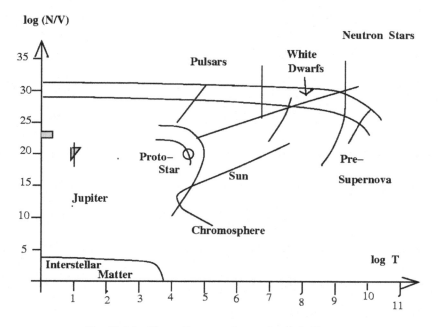

Fig. 11.1.1   Phase diagram of astrophysical objects.

of Figure 11.1.1 while ordinary stars lie on the lower region. Pulsars are actually cold neutron stars. Of these astrophysical objects, stars form a very important component, not only because they are the beacons for studying the depth of space, but also because all the elements of the periodic table are created during the thermonuclear processes occurring within the stars. In the study of astrophysics use is made of different branches of physics, *classical mechanics, fluid mechanics, atomic and nuclear physics, gravitational theory,* and, of course, *statistical physics*, since astrophysical objects form a many-body system. Though in the astronomical time scale they are *not* in thermodynamic equilibrium, in the terrestrial time scale of measurement the system is assumed to be in statistical equilibrium and we shall use the results that we have obtained in § 9 and § 6.

11.2   Stars: Stability and Evolution

A star may be defined as a *heavenly body which is opaque to its own radiation.* Stability of a star of mass $\mathcal{M}$ and radius R in hydrostatic equilibrium is established through interplay of its gravitational energy and its internal energy, manifested by internal pressure. A star of radius R and mass $\mathcal{M}$ with a uniform density will be in hydrostatic equilibrium when its total energy

$$E_{\text{total}} = E_{\text{grav}} + E_{\text{int}}$$
$$= -\frac{3}{5}\frac{G\mathcal{M}^2}{R} + E_{\text{int}} \tag{11.2.1}$$

is minimum, i.e. $\frac{\partial E_{\text{total}}}{\partial R} = 0$ as the first condition. Since

$$\frac{\partial E_{\text{grav}}}{\partial R} = -\frac{E_{\text{grav}}}{R}, \tag{11.2.2}$$

and

$$\frac{\partial E_{\text{int}}}{\partial R} = \frac{\partial E_{\text{int}}}{\partial V}\frac{\partial V}{\partial R} = -\frac{3PV}{R} \tag{11.2.3}$$

we obtain the **Virial theorem of stellar stability**

$$E_{\text{grav}} + 3PV = -R\frac{\partial E_{\text{total}}}{\partial R} = 0. \tag{11.2.4}$$

The pressure needed for the stability of a star must be supplied by degeneracy pressure of the fermions within the star as calculated in § 9.7. Since the degeneracy pressure depends on particle density and temperature, these should be high enough to give stability to a star of given mass $\mathcal{M}$ and radius R. For E_{total} to be minimum, a second condition

$$
\begin{aligned}
\frac{\partial^2 E_{\text{total}}}{\partial R^2} &= \frac{2E_{\text{grav}}}{R^2} - \frac{6PV}{R^2} - \frac{3V}{R}\frac{\partial P}{\partial R} \\
&= -\frac{9PV}{R^2}\left[\frac{4}{3} + \frac{V}{P}\frac{\partial P}{\partial V}\right] > 0.
\end{aligned}
\tag{11.2.5}
$$

should also be satisfied. This gives us the constraint for the *compressibility*

$$
\Gamma = -\frac{V}{P}\left(\frac{\partial P}{\partial V}\right) \geq \frac{4}{3}.
\tag{11.2.6}
$$

For a non-relativistic gas $\Gamma < \frac{5}{3}$, while for a relativistic gas $\Gamma < \frac{4}{3}$. All dissociations

$$
H_2 \leftrightarrow 2H, \quad H \leftrightarrow H^+ + e^-,
\tag{11.2.7}
$$

$$
e^+ + e^- \leftrightarrow \gamma + \gamma, \quad {}^{56}Fe \leftrightarrow 13\ {}^4He + 4n
\tag{11.2.8}
$$

correspond to $\Gamma < \frac{4}{3}$; so these processes cannot occur within a stable stellar interior.

We now consider the *evolution process* of a star. The absolute surface luminosity of a star is plotted in the **Hertzsprung-Russell diagram** as

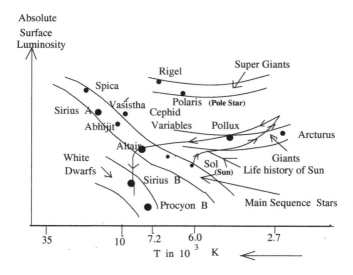

Fig. 11.2.1   Hertzsprung-Russell diagram.

a function of its temperature (Figure 11.2.1); it is to be noted that temperature increases towards the left. We have now plotted the positions of some of the important stars in the **Hertzsprung-Russell diagram**. As an aside, we give in Table 11.2.1 the names used by the Hindu Astronomers for some of these stars:

Table 11.2.1   Equivalent Names of some stars in Hindu Astronomy.

Star Names	Indian Names
Arcturus	Svātī
Altair	Śravanā
Polaris (Pole Star)	Dhruvatārā
Pollux	Punarvasu
Sirius	Lubdhaka
Sol (Sun)	Sūrya
Spica	Citrā

We have written the Indian names in Roman letters of alphabet using the rule suggestested by the International Alphabet for Sanskrit Transliteration (IAST) and is given in Table 14.10.1 in the Mathematical Appendix.

About 80% of all the stars lie on the **Main Sequence** shown in the Hertzsprung-Russell (HR) diagram (Figure 11.2.1). The heavier a star is, the higher it is situated on the HR diagram. In astrophysical literature, mass $\mathcal{M}_\odot = 1.99 \times 10^{33}$ g, and radius $R_\odot = 6.96 \times 10^{10}$ cm of the sun are taken as the units in which stellar mass and radius are measured. Stars in other groupings on HR diagram are characterized by

$$100 < \frac{R}{R_\odot} < 1000 : \text{Super Giants}$$

$$10 < \frac{R}{R_\odot} < 100 : \text{Giants}$$

$$0.01 < \frac{R}{R_\odot} < 0.2 : \text{White Dwarfs}$$

At the beginning of a star's life, when only hydrogen is present within it, protons burn to form helium. For light stars the relevant nuclear process is the **proton cycle**

$$^1H(p, e^-)\,^2H(p, \gamma)\,^3He(He^3, 2p)\,^4He,$$

while for more massive stars the carbon-nitrogen-oxygen-cycle (**CNO tricycle**)

$$^{12}C(p,\gamma)^{13}N(e^+,\nu)^{13}C(p,\gamma)^{14}N(p,\gamma)^{15}O(e^+,\nu)^{15}N(p,\alpha)^{12}C+26.731\text{MeV}$$

produces energy within the star. When all the hydrogen within the star is exhausted, **helium burning process**

$$^4He(2\alpha,\gamma)^{12}C(\alpha,\gamma)^{18}(\alpha,\gamma)^{20}Ne + 7.275\text{MeV} + 7.162\text{MeV}$$

commences. Successive nuclear reaction cycles, among others,

$$\text{Carbon burning}: {}^{12}C + 12C \to {}^{24}Mg^* \to {}^{23}Mg + n - 2.62\text{MeV}$$
$$\to {}^{20}Ne + \alpha + 4.62\text{MeV}$$
$$\to {}^{23}Na + p + 2.24\text{MeV}$$
$$\text{Neon burning}: {}^{20}Ne + {}^{20}Ne \to {}^{16}O + {}^{24}Mg + 4.59\text{MeV}$$
$$\text{Oxygen burning}: {}^{16}O + {}^{16}O \to {}^{31}S^* \to {}^{31}S + n + 1.45\text{MeV}$$
$$\to {}^{31}P + p + 7.68\text{meV}$$
$$\to {}^{30}P + d - 2.41\text{meV}$$
$$\to {}^{21}Si + \alpha + 9.59\text{meV}$$

and finally the Silicon 'melting' process

$$^{28}Si(\gamma,\alpha)^{24}Mg(\gamma,\alpha)^{20}Ne(\gamma,\alpha)^{16}O(\gamma,\alpha)^{12}C(\gamma,2\alpha)\alpha$$

are the principal thermonuclear reactions that keep the star living. It should be remembered that all these and other reactions occur at the core of the star. The photons and neutrinos escape the star carrying away energies. Transfer of energy from the stellar core is radiative in the layer immediately adjacent to the core and is convective in the outermost layer of the star. As long as there are nuclear fuels available for burning the pressure generated thereby will balance the gravitational pressure and the star is stable.

As successive thermonuclear processes in the interior of a star progress the star starts moving from its position on the Main Sequence band. The projected motion of the Sun is shown in Figure 11.2.1. All Main Sequence stars have similar evolution path. For a heavy star the speed of the evolution is much faster than that for a less massive star. Stars of masses similar to that of the sun end as **White Dwarfs** while more massive stars end as **Neutron Stars**, and the heaviest ones become **Black Holes**.

11.3   High Temperature Dense Matter

A study of highly dense matter at high temperature is of interest for under-standing the processes at stellar core. The case of hydrogen as prototype is shown in Figure 11.3.1. Some portion of the figure is from laboratory

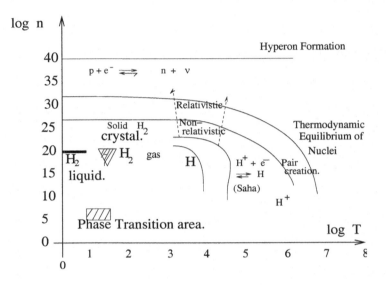

Fig. 11.3.1   Phase diagram of Hydrogen.

experiments, while other portions are from theoretical calculations. Notice-able similarity is observed with Figure 11.1.1. The reason is that hydrogen forms by far the most abundant element in the universe.

The transition from neutral to ionized hydrogen is obtained from Saha formula, discussed in §6.5. Ground state of hydrogen does not exist when the potential of the nearest neighbors depresses the continuum to the level of the bound electron. This gives a natural distance $a = a_B$, where a_B is the Bohr radius and the number density of hydrogen $n_H = 10^{23.4}$. The bound-ary between degenerate and non-degenerate region is non-relativistically given by $n_H \approx 10^{16.2} T^{3/2}$. In the relativistic case the boundary is de-termined by pair creation process with the number density $n_H \approx 10^{11} T^2$. Solidification occurs with melting temperature of an atom of atomic number Z and mass density ρ given by $T_{\text{melting}} \approx 3 \times 10^5 (\frac{\rho}{10^6})^{\frac{1}{3}} Z^{\frac{2}{3}}$ K. Correspond-ing number density for hydrogen is $n_H \approx 10^{12.3} T^3$. At very high densities neutrons are in equilibrium with protons, since beta decay is forbidden for completely filled Fermi spheres. With energy of 1.29 MeV this gives the number density of $n_H \approx 10^{30.93}$, where neutrinos are completely ignored. Very high temperatures are dominated by radiations and pair production is the dominant process.

We are now in a position to describe successive processes within the core as a star evolves. When volume per atom $\ll$ size of an atom, atoms loose their individuality and a highly compressed plasma of e^- and nuclei is formed. If T is not very high, electrons become a degenerate Fermi gas. If N_e = number of electrons, then

$$a = \text{separation between } e^- \text{ \& nucleus} \propto \left(\frac{ZV}{N_e}\right)^{\frac{1}{3}}.$$

Here we have used number of atom $= \frac{N_e}{Z}$ and atomic size $= \frac{V}{(N_e/Z)}$. In this case

$$\text{Coulomb Energy} = \frac{Ze^2}{a} \ll \mu_0, \text{ Mean Kinetic Energy}$$

and using the results of § 9.7 we get for the electron density

$$n_e \equiv \frac{N_e}{V} \gg Z^2 \left(\frac{m_e e^2}{\hbar^2}\right)^3, \tag{11.3.1}$$

where m_e is the mass of electron. As $\frac{N_e}{V}$ increases the system behaves more and more like a degenerate Fermi gas of electrons and the presence of protons can be neglected as they are far from degeneracy. The corresponding inequality for the density of total mass is

$$\rho = n_e m' \gg \left(\frac{m_e e^2}{\hbar^2}\right)^3 m' Z^2 \sim 20 Z^2 \text{g/cm}^3, \tag{11.3.2}$$

where m' is the mean atomic mass. The corresponding degeneracy temperature is $T_g \sim 10^6 Z^{\frac{4}{3}} K$. In the non-relativistic case when $p_F \ll m_e c$, the pressure satisfies

$$P = \frac{(3\pi^2)^{\frac{2}{3}}}{5} \frac{\hbar^2}{m_e} \left(\frac{\rho}{m'}\right)^{\frac{5}{3}} \gg 5 \times 10^8 Z^{\frac{10}{3}} \text{bar}. \tag{11.3.3}$$

In case $\rho \ll 2 \times 10^6 \text{g/cm}^3$, $P \ll 10^{17}$ bar, where 1 bar $= 1.01325 \times 10^5$ Pa $= 1.01325 \times 10^5$ Newton/m^2. In the relativistic case the corresponding expression is $P = \frac{1}{4} \left(3\pi^2\right)^{\frac{1}{3}} \hbar c \left(\frac{\rho}{m'}\right)^{\frac{4}{3}}$.

With further increase of density nuclear reaction processes begin and the most important one for our purpose is the β-capture process, $p + e^- \rightleftharpoons n + \nu$

for a single proton and $A_Z + e^- \rightleftharpoons A_{Z-1} + \nu$ for a nucleus with mass number A and atomic number Z. Neutrinos leave the system carrying away energy and the system cools down and we can use zero-temperature statistical physics of § 9.7. Binding energy $\epsilon(A, Z)$ of the (A, Z) nucleus is connected with its chemical potential $\mu(A, Z)$ by the relation $\epsilon(A, Z) = -\mu(A, Z)$. Condition of chemical reaction equilibrium of § 6.1 gives us (since $\mu(\nu) = 0$)

$$\mu_e(n_e) = \mu(A, Z - 1) - \mu(A, Z)$$
$$= \epsilon(A, Z) - \epsilon(A, Z - 1) = \Delta. \tag{11.3.4}$$

Results of § 9.2 allows us to write for the number density of electron and therefore for its pressure as

$$n_e = \frac{\Delta^3}{3\pi^2 (\hbar c)^3} = \text{constant}, \tag{11.3.5}$$

$$P = \frac{\Delta^4}{12\pi^2 (\hbar c)^3} = \text{constant}. \tag{11.3.6}$$

As the density increases further, more and more β-capture occurs. Nuclei will then be so neutron-rich that they break up and individual nuclei lose their individuality. At $\rho \approx 3 \times 10^{11} \frac{g}{cm^3}$, when $P \approx 10^{24}$ bar neutrons become more numerous than electrons. When density increases by another factor of 10, neutrons predominate and everything else is neglected. Pressure now satisfy the *polytropic relation*

$$P = \frac{\left(3\pi^2\right)^{\frac{2}{3}}}{5} \frac{\hbar^2}{m_{\text{neutron}}^{\frac{8}{3}}} \rho^{\frac{5}{3}} = 5.5 \times 10^3 \rho^{\frac{5}{3}} \text{ bar}. \tag{11.3.7}$$

At $\rho \gg 6 \times 10^{15} \frac{g}{cm^3}$, neutrons become extremely relativistic and pressure becomes

$$P = \frac{\left(3\pi^2\right)^{\frac{1}{3}}}{4} \hbar c \left(\frac{\rho}{m_{\text{neutron}}}\right)^{\frac{4}{3}} = 1.2 \times 10^9 \rho^{\frac{4}{3}} \text{ bar}. \tag{11.3.8}$$

At this stage of development, *strong nuclear force* comes into play.

11.4   Neutron Stars and Black Holes

We have mentioned in § 11.2 that ordinary stars of mass of the order of solar mass $\mathcal{M}_\odot$ become white dwarfs when all their nuclear fuel is exhausted. Stars having the mass of their core exceeding $1.44 \mathcal{M}_\odot$ explode as novae and supernovae when their nuclear fuel is burnt out. This condition on the mass of the core is called the **Chandrasekhar limit**. In a supernova explosion the outer parts of the star is ripped open and the material there gets mixed with interstellar medium, later to form new stars. The core ends up as a **Neutron Star** (NS) or a **Black Hole** (BH). Typically a Neutron Star has mass between $0.1\mathcal{M}_\odot$ and $3.0\mathcal{M}_\odot$, while Black Holes in X-ray binaries have mass in the range $5\mathcal{M}_\odot$ $-20\mathcal{M}_\odot$ and Black Holes in Galactic Nuclei have masses in the range $10^6\mathcal{M}_\odot-10^{9.5}\mathcal{M}_\odot$. **Pulsars** and **Compact X-Ray Sources** are the most likely places in the universe where Neutron Stars and Black Holes are expected to be found. About 150 Pulsars and 10 X-Ray Sources have been identified.

With radius $R_{NS} \approx 10^6$ cm the mass density within a Neutron Star is well above the nuclear matter density $\rho_{\text{nuclar}} \approx 2.8 \times 10^{14}$ $\frac{\text{g}}{\text{cm}^3}$. A neutron Star is thus a *giant nucleus* with Mass Number $A \sim 10^{57}$ nucleons. All neutron stars have been found to have large *surface magnetic field* $B \sim 10^{12}$ Gauss. Neutron Stars have been ascribed internal structures. A possible structure is shown in Figure 11.4.1.

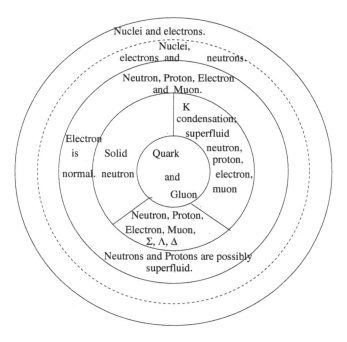

Fig. 11.4.1 Possible structure of Neutron Stars.

A Black Hole is assigned two parameters, its *mass* $\mathcal{M}_{BH}$ and its *spin* a_S. The most remarkable property of a Black Hole is the existence of **Event Horizon** of radius R_S. Through the Event Horizon particles and radiations can fall *on* the Black Hole, but *no* particle or radiation *from* the Black Hole can come out. For a spinning Black Hole , $R_S = \frac{2G\mathcal{M}_{BH}}{c^2} = 2.95 \times 10^5 \left(\frac{\mathcal{M}_{BH}}{\mathcal{M}_\odot} \right)$ cm. For a non-spinning Black Hole, $R_S \longrightarrow \frac{G\mathcal{M}_{BH}}{c^2}$, as $a_S \longrightarrow 1$. For our own Milky Way galaxy the Black Hole at its nucleus has a mass $\mathcal{M}_{BH} \approx (3.7 \pm 0.2) \times 10^6 \, \mathcal{M}_\odot$ and radius $\approx 2 \times 10^{15}$ cm.

11.5   Problems

Problem 11.1.   Prove Equation 11.2.5.

Problem 11.2.   Prove Equation 11.3.3.

Problem 11.3.   Prove Equation 11.3.4.

Problem 11.4.   Prove Equation 11.3.5.

Problem 11.5.   Prove Equation 11.3.6.

Problem 11.6.   Prove Equation 11.3.7.

Problem 11.7.   Prove Equation 11.3.8.

Chapter 12

Phase Transitions

12.1 Systematics of Phase Transitions

Transition of a system from one phase to another occurs in all natural systems. The phase diagrams given in Figures 12.1.1 & 12.1.2 show phase transition for water and a magnetic system. These are two important prototypes of phase transiions observed in nature. While for water there is

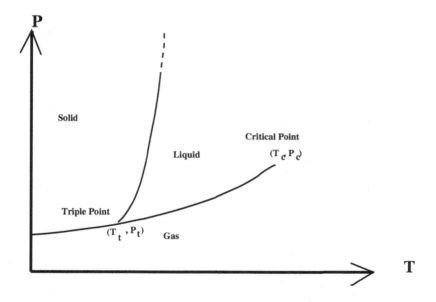

Fig. 12.1.1   Phase diagram of water.

always a *Latent Heat* of transition, in magnetic transitions from the paramagnetic to the ferromagnetic phase there is no Latent Heat. This property distinguishes these two prototypes. The *(P-T)* liquid-gas phase co-existence curve for water terminates at the Critical Point just like the paramagnetic-feromagnetic phase co-existence curve *(H-T)* for the magnetic system at the Curie Point. The solid-liquid phase co-existence curve, however, has not yet shown any end point for all systems having water as a prototype. It is to be pointed out that in many organic liquids the **triple point** is replaced by a '*triple triangle*' within which all the three phases coexist, their relative masses being given by Maxwell's triangle law. Magnetic Systems, on the

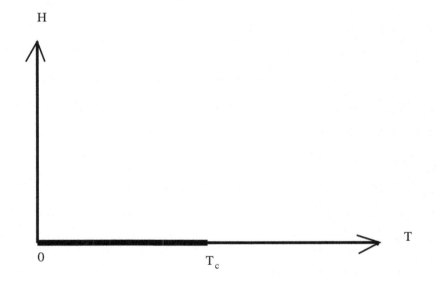

Fig. 12.1.2   Phase Diagram of magnetic systems.

other hand, show a rich plethora of interesting phase transitions.

In this connection we show in Figure 12.1.3 the phase diagram for a transition from superconducting phase to normal phase.

Another way of depicting these phase diagrams is plotting them in density (ρ)-temperature (T) space as in Figure 12.1.4 instead of in the pressure (P)-temperature (T) space for water and the magnetization (M)-temperature (T) space as in Figure 12.1.5 instead of the magnetic field (H)-temperature (T) space for magnetic systems.

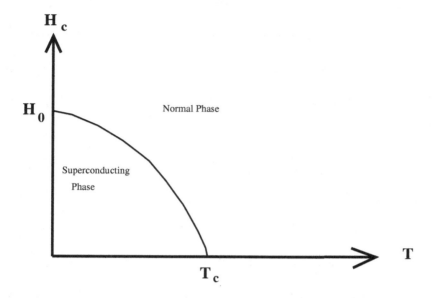

Fig. 12.1.3   Critical field versus temperature curve for superconducting systems.

12.2   Ehrenfest's Classification of Phase Transitions

Early in the twentieth century Paul Ehrenfest gave a classification scheme for different classes of phase transitions. In spite of later and modern schemes of classifications, physicists still use Ehrenfest's terminology. This scheme has the advantage that it refers to experimentally measured quantities like Latent Heat and Specific Heat.

This classification scheme depends on the continuity properties of the Gibbs' potential G and its derivatives of different order with respect to temperature T.

If $G^{(r)} = \frac{\partial^r G}{\partial T^r}$ so that $G^{(0)}$ is the Gibbs' potential itself, then according to this classification scheme for an *n-th order* transition from phase I to phase II

$$G_I^{(k)} = G_{II}^{(k)}, \quad \text{for all } 0 \leq k \leq n-1, \tag{12.2.1}$$

$$G_I^{(n)} \neq G_{II}^{(n)}. \tag{12.2.2}$$

This means that the n-th order derivative of the Gibbs' potential is discontinuous while all the lower order derivatives including the function itself are continuous.

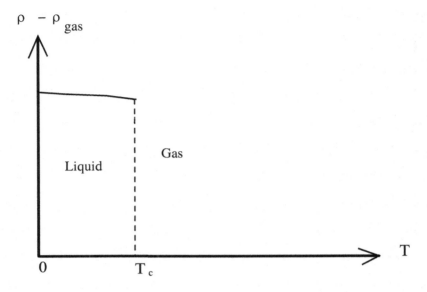

Fig. 12.1.4   Density versus temperature curve of water.

For a **first order transition**, n=1 and *entropy is discontinuous*, thus showing the presence of *latent heat of transition*; and for a **second order transition**, n=2 and *specific heat* is discontinuous. As a matter of fact latent heat of transition is a function of temperature and the nature of this dependence is shown in Figure 12.2.1.

In Ehrenfest's classification derivatives of Gibbs' potential with respect to other thermodynamic variables like pressure have no role.

12.3   Order Parameter, Continuous and Discontinuous Transitions

Landau shifted the attention from changes in thermodynamic variables like entropy and specific heat to changes in properties like density and magnetization as the system undergoes a phase transition. The two prototypes, liquid-gas transition and ferromagnetic-paramagnetic transition, played a significant role in this shift of emphasis. Density and magnetization are not merely thermodynamic variables and their changes are connected with application of external force fields. Landau termed these variables **order parameters**. For liquid-gas transition, the difference of density $\rho - \rho_g$ is

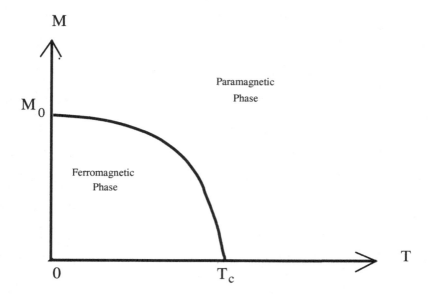

Fig. 12.1.5   Magnetization curve.

the order parameter; for magnetic systems, magnetization **M** is the order parameter; for ferroelectric systems, dielectric polarization **P** is the order parameter. Order parameters in these cases are all directly measurable quantities. But this need not always be the case. The classic case of order-disorder transition in binary alloys like *Cu-Zn* belongs to this category. In Figure 12.3.1 we have shown the equilibrium positions of *Cu* and *Zn* in the solid. We now denote by $W_{\alpha:\beta}$ the probability that the atom of type α occupies the normal position of the atom of type β, where α and β are either *Cu* or *Zn* and define the *order parameter* as

$$\eta = \frac{W_{Cu:Cu} - W_{Zn:Cu}}{W_{Cu:Cu} + W_{Zn:Cu}}. \tag{12.3.1}$$

In the perfectly ordered phase when all the atoms occupy their equilibrium positions, $W_{\alpha:\beta}$ is 1 if $\beta = \alpha$ and is 0 if $\beta \neq \alpha$, and so $\eta = 1$; in the perfectly disordered phase when any atom occupies any position with equal probability, $W_{\alpha:\beta}$ is always 0.5, and thus $\eta = 0$. This order parameter is surely *not* directly measurable by experiments. Similarly for superconducting materials the order parameter is defined as $\eta = \langle C_{\mathbf{k}\uparrow}^{\dagger} C_{\mathbf{k}\downarrow}^{\dagger} \rangle$. Here $C^{\dagger}$ denotes the corresponding *creation* operator. This is again a case where

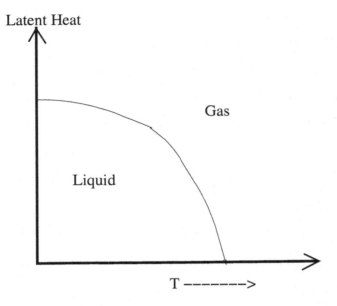

Fig. 12.2.1   Dependence of latent heat on temperature.

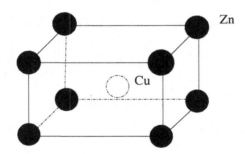

Fig. 12.3.1   Equilibrium positions of *Cu* and *Zn* in a solid.

the order parameter cannot be directly measured. Moreover in ferroelectric transitions, together with development of non-zero electric polarization, the crystal symmetry changes from cubic to tetragonal class indicating genera-tion of elastic strain tensor. When there are more than one order parameter, one of them is the **principal order parameter**. In the ferroelectric case the dielectric polarization is the principal order paramater and the elastic strain is the **secondary order parameter**. There is unfortunately no algorithm for choosing which one is the principal order parameter. It is more or less a case of physical intuition. The choice of the correct order parameter itself is also the investigator's physical acumen.

12.4 Landau's Theory of Continuous Phase Transitions

Landau classified phase transitions according to the behaviour of the order parameter η shown in Figures 12.4.1 & 12.4.2. In the case of **discontin-**

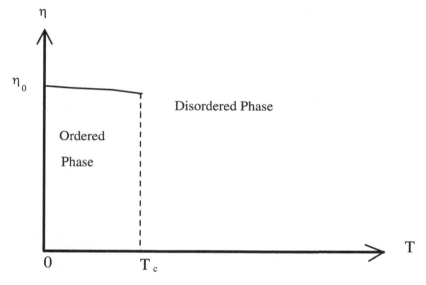

Fig. 12.4.1 Variation of order parameter η as a function of temperature in the case of a discontinuous phase transition.

uous phase transition the order parameter η changes discontinuously at the transition temperature T_c, while for a **continuous phase transition** η changes continuously and vanishes at T_c. In the case of *liquid-gas* transition the density $\rho - \rho_{gas}$ definitely satisfies this criterion as does magnetization **M** in the case of magnetic transition.

$$\eta = 0, \quad \text{in the higher symmetry disordered phase,} \qquad (12.4.1)$$

$$\eta \neq 0, \quad \text{in the lower symmetry ordered phase.} \qquad (12.4.2)$$

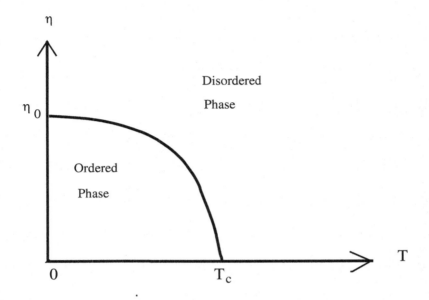

Fig. 12.4.2   Variation of order Parameter η as a function of temperature in the case of a continuous phase transition.

Order parameters for superconducting and *Cu-Zn* like order-disorder transitions described in Section 12.3 also have this characteristics.

It should also be noted that in a system there may be more than one phase transition point at different temperature like the case of $\alpha - Fe_2O_3$ which is paramagnetic above 950K and two antiferromagnetic ordering, one below 250K and the other in the range 250K–950K.

Landau's analysis is based on analytic expansion of the Free Energy $F(p,T,\eta)$ in powers of η:

$$F(p, T, \eta) = F_0(p, T) + \alpha(p, T)\eta + A(p, T)\eta^2 + C(p, T)\eta^3 + B(p, T)\eta^4.$$
$$(12.4.3)$$

This formula is valid only when there is no external field. In Landau's theory of *second order phase transition* the expansion co-efficients $\alpha(p,T)$, $A(p,T)$, $C(p,T)$ and $B(p,T)$ are analytic functions of the arguments. Landau's theory crucially depends on the symmetry argument that the two phases with $\eta = 0$ and $\eta \neq 0$ have different symmetries. From this Landau concluded, the details of which we omit, that $\alpha(p, T) \equiv 0$ as an identity. Next $B(p,T_c)$ has to be strictly positive; otherwise Free energy will decrease indefinitely for large positive and negative values of η thus precluding any minimum of F. Thus $B(p,T) = B(p) > 0$ on the phase co-existence curve. We now plot in Figure 12.4.3 the Free energy $F(p,T,\eta)$- $F(p,T,\eta = 0)$ as

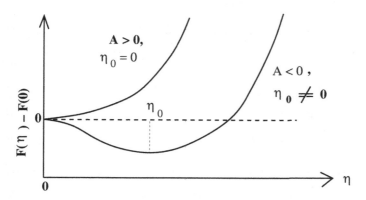

Fig. 12.4.3 Free energy $F(p,T,\eta)$-$F(p,T,\eta=0)$ as a function of the order parameter η.

a function of η. It is evident that with $A(p,T) > 0$, F has a minimun at $\eta = 0$; while for $A(p,T) < 0$ the minimum is at $\eta \neq 0$. Thus at the transition point T_c, $A(p,T_c) = 0$. We must remember that this is an equation and not a identity. If moreover $C(p,T_c) = 0$ is true as an equation then solutions of two equations give isolated transition points. Such transitions exist in nature but are not describable by Landau's theory. Thus in this theory we must have the identity $C(p,T) \equiv 0$. The Landau Free Energy thus have the form:

$$F(p,T,\eta) = F_0(p,T) + A(p,T)\eta^2 + B(p)\eta^4. \qquad (12.4.4)$$

Landau made the further *assumption*

$$A(p,T) = a(p)(T - T_c), \quad \text{with} \quad a(p) > 0. \qquad (12.4.5)$$

12.5 Continuity of Entropy and Discontinuity of Specific Heat

We now obtain expressions for entropy and specific heat in the Landau theory. On the phase co-existence curve where the two phases are in equilibrium the value of the order parameter η_0 is obtained from the equation

$$\frac{\partial F}{\partial \eta_0} = 0, \tag{12.5.1}$$

leading to

$$\eta_0 (A + 2B\eta_0^2) = 0. \tag{12.5.2}$$

When $T > T_c$, the only possible solution is $\eta_0 = 0$ corresponding to the disordered phase. On the other hand when $T < T_c$ of the two solutions of Equation 12.5.2 $\eta_0 = 0$ corresponds to the maximum value of the Free Energy, and thus it does not correspond to thermodynamic equilibrium. On the other hand the other solution

$$\eta_0 = \sqrt{-A(p,T)/2B} \tag{12.5.3}$$

corresponds to the thermodynamical equilibrium state with minimum Free Energy F. We thus write

$$\eta_0 = \sqrt{\frac{a(p)}{2B}} |t|^{\frac{1}{2}}, \quad \text{with} \tag{12.5.4}$$

$$t = T - T_c \quad \text{and} \tag{12.5.5}$$

$$\eta_0 \propto |t|^{\frac{1}{2}}. \tag{12.5.6}$$

We now investigate the behaviour of entropy, latent heat and specific heat at constant pressure in the case of a continuous phase transition. Remembering the thermodynamic definitions of entropy $S = -\frac{\partial F}{\partial T}$, and specific heat at constant pressure $C_p = T \left(\frac{\partial S}{\partial T} \right)_p$ we obtain

$$S = S_0 + \frac{a(p)^2}{2B} (T - T_c), \tag{12.5.7}$$

$$C_p = C_{p0} + \frac{a(p)^2}{2B} T. \tag{12.5.8}$$

At the phase transition temperature $T = T_c$ we obtain $S = S_0$ indicating continuity of entropy and thus absence of any latent heat of transition. For the specific heat at constant pressure we get $C_p = C_{p0} + \frac{a(p)^2}{2B} T_c$ showing that there is a discontinuity across the phase co-existence curve. This discontinuity is represented as

$$C \propto |t|^0. \tag{12.5.9}$$

Similar discontinuity exists also in the behaviour of compressibility, thermal expansion coefficient and specific heat at constant volume.

Behaviour of latent heat and specific heat indicates that Ehrenfest's second order phase transition is a particular case of Landau's continuous phase transition.

12.6 Generalized Susceptibility

Analogous to the magnetic susceptibility of natural magnetic systems, in every case of continuous phase transition there is a temperature dependent **generalized susceptibility**. This generalized susceptibility controls the response of the order parameter to external force field. To obtain this temperature dependence we have to modify the zero-field Landau Free Energy F of §12.4 by introducing a finite force field h and define the Gibbs' potential

$$G(p, T, \eta) = G_0(p, T) + at\eta^2 + B\eta^4 - \eta hV, \quad \text{with} \quad (12.6.1)$$

$$t = T - T_c. \quad (12.6.2)$$

The external field term with the sample volume V has been written in analogy to that in the prototype magnetic case. The equilibrium value of the *Order Parameter* η_0 is obtained by solving the minimality condition of G given in Equation 12.6.1, $\left(\frac{\partial G}{\partial \eta}\right)_h = 0$ at $\eta = \eta_0$:

$$2at\eta_0 + 4B\eta_0^3 = hV. \quad (12.6.3)$$

Equation 12.6.3 being a cubic equation in η_0 can, of course, be solved algebraically; but in that process we shall miss the physical nature of the solution. In the graphical method of solving the equation $f_1(\eta) = f_2(\eta)$ we plot both $f_1(\eta)$ and $f_2(\eta)$ as functions of η and the points of intersection of the curves give us the solutions (η_0). We shall take $f_1(\eta) = 2at\eta + 4B\eta^3$ and $f_2(\eta) = hV$. The function $f_2(\eta)$ is a straight line parallel to the abscissa and depending on the value of h at various distance from the abscissa. Varying the value of h the points of intersection of the two curves and thus the values of η_0 are obtained as functions of h.

In the disordered phase when $t > 0$ both $f_1(\eta)$ and $\eta_0(h)$ are monotonously increasing functions of their respective arguments and are not of much interest. This actually corresponds to the paramagnetic phase of magnetic systems.

In the ordered phase, $t < 0$ and we have $t = -|t|$. The function $f_1(\eta)$ now takes the form $-2a|t|\eta + 4B\eta^3$. In this form the polarity of the in-

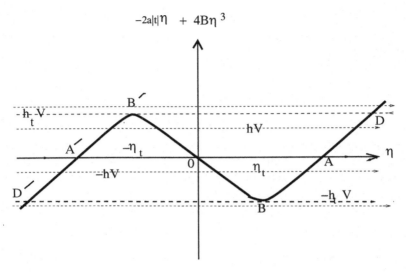

Fig. 12.6.1   Schematic plot of $f_1(\eta) = -2a|t|\eta + 4B\eta^3$ and $f_2(\eta) = hV$ as functions of η. $f_2(\eta)$ has been plotted for different values of h.

dividual terms has been explicitly manifested. In Figure 12.6.1 we have schematically plotted both $f_1(\eta)$ and $f_2(\eta)$ for different values of the external force field h. We note that when $h > h_t$ there is only one positive root η_0; while for $0 \le h < h_t$ there are three roots, one positive and two negative. When $h = h_t$ the two negative roots coalescese to one double root of negative value. Analogous situation exists when $h < 0$. Extracting data from Figure 12.6.1, we again schematically plot in Figure 12.6.2 the curve of η_0 as a function of h.

We shall remember that on the curve $DABOB'A'D'$, G having an extremum,

$$\frac{\partial G}{\partial \eta_0} = f(\eta_0, h) \equiv 2at\eta_0 + 4B\eta_0^3 - hV = 0. \tag{12.6.4}$$

This immediately gives us

$$\frac{df}{dh} \equiv \frac{\partial f}{\partial h} + \frac{\partial \eta_0}{\partial h}\frac{\partial f}{\partial \eta_0} = 0, \tag{12.6.5}$$

$$-V + \frac{\partial^2 G}{\partial \eta_0^2}\frac{\partial \eta_0}{\partial h} = 0. \tag{12.6.6}$$

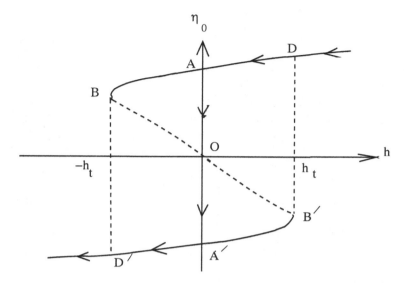

Fig. 12.6.2 Schematic plot of η_0 as a function of h. The points A, A', B, B', D, D' and O correspond to the same points in Figure 12.6.1.

On BB', $\frac{\partial \eta_0}{\partial h} < 0$, and so $\frac{\partial^2 G}{\partial \eta_0^2} < 0$. Thus G is maximum in this branch and the system is thermodynamically unstable. G is minimum on the branches AB and $A'D'$ as also on AD and $A'B'$; but $G_{AB} > G_{A'D'}$ and $G_{A'B'} > G_{AD}$. Thus the branches AB and $A'B'$ are metastable. As the external force field is decreased from a high value the system follows the path $DAOA'D'$. The jump from A to A' will occur because in every finite system at finite temperatures, fluctuations are always present. This jump is like the one in $P - V$ characteristics of liquid-vapour mixture below the critical temperature and follows the same **Maxwell's lever rule**.

We can now define **Generalized susceptibility** as

$$\chi \overset{\text{def}}{=} \frac{\partial \eta_0}{\partial h} = V / \left(\frac{\partial^2 G}{\partial \eta_0^2} \right). \qquad (12.6.7)$$

Using the values of η_0 in the disordered phase ($\eta_0 = 0$) and the ordered phase (Equation 12.5.4) we get

$$\chi = \frac{V}{2at}, \quad \text{disordered phase with } t > 0, \ \eta_0 = 0, \qquad (12.6.8)$$

$$\chi = \frac{V}{4a|t|}, \quad \text{ordered phase with } t < 0, \ \eta_0 \neq 0. \qquad (12.6.9)$$

In both the phases χ obeys a power-law behaviour

$$\chi \propto |t|^{-1}, \tag{12.6.10}$$

but with two different proportionality constants in the two phases.

We also have a power law dependence of the *critical field* h_t when two values of the equilibrium η_0 coalesce to a single η_t. Using the extremal nature of the function on the left hand side of the Equation 12.6.3 at η_t and h_t we get

$$h_t = \frac{\left(\frac{2}{3}a|t|\right)^{\frac{3}{2}}}{VB^{\frac{1}{2}}}, \tag{12.6.11}$$

$$\eta_t = \sqrt{\frac{a|t|}{6B}}. \tag{12.6.12}$$

Joining the last two equations we get a power law relation

$$\eta_t = \left(\frac{h_t V}{8B}\right)^{\frac{1}{3}}, \tag{12.6.13}$$

$$\eta_t \propto h_t^{\frac{1}{3}}. \tag{12.6.14}$$

In Landau's theory no attempt has been made to explain phase transition from the dynamics of the system like using a hamiltonian. Only a Free Energy, which is an averaged out quantity, has been used. It is thus only a **Mean Field Theory**. Moreover, no allowance has been made for fluctuations of the order parameter. Generally, very close to the critical points these fluctuations grow in importance and Landau theory breaks down. However, there are cases like structural phase transitions, where these fluctuations die down and Landau thery works even up to the critical point. At least for a wide range of temperature Landau results are valid.

12.7   Critical Exponents and Fluctuations of Order Parameter

In Equations 12.5.6, 12.5.9, 12.6.10 and 12.6.14 we have obtained power-law dependence of different physical quantities. Though these have been derived there only for Landau model, these power-law relations are true in general:

$$\eta \propto |t|^{\beta}, \tag{12.7.1}$$

$$C \propto |t|^{-\alpha}, \tag{12.7.2}$$

$$\eta \propto h^{\frac{1}{\delta}}, \tag{12.7.3}$$

$$\chi \propto |t|^{-\gamma}, \tag{12.7.4}$$

where α, β, γ and δ are called the **critical exponents**. In Landau theory these critical exponents have the values

$$\alpha = 0, \ \beta = \frac{1}{2}, \ \gamma = 1, \text{ and } \delta = 3. \tag{12.7.5}$$

These critical exponents are not really independent. A number of inequalities involving them have been derived

$$\text{Rushbrook}: \alpha + 2\beta + \gamma \geq 2, \tag{12.7.6}$$

$$\text{Griffith}: \quad \beta \cdot \delta \quad \geq \quad \beta + \gamma. \tag{12.7.7}$$

Later on it was found that the strict equalities are actually true

$$\text{Essam} - \text{Fischer}: \alpha + 2\beta + \gamma = 2, \tag{12.7.8}$$

$$\text{Widom}: \quad \beta \cdot \delta \quad = \quad \beta + \gamma. \tag{12.7.9}$$

Derivation of these results uses the analyticity property of the thermodynamic functions away from critical points. The Landau values of the critical exponents satisfy all of them.

In the Landau analysis the order parameter was assumed to be the same all over the sample and thus fluctuation of the order parameter was ignored. Very near to the critical point these fluctuations cannot be ignored and these are accounted for by introducing an extra term involving space derivative of the order parameter in the Landau Free Energy

$$G = G_0 + at\eta^2 + b\eta^4 - h\eta V + \int g \left(\nabla \eta\right)^2 d\mathbf{r} \tag{12.7.10}$$

and we define **correlation function** of fluctuations of order parameters at space points $\mathbf{r}_1$ and $\mathbf{r}_2$ with $\mathbf{r} = \mathbf{r}_1 - \mathbf{r}_2$

$$C(\mathbf{r}) = \langle \Delta\eta(\mathbf{r}_1)\Delta\eta(\mathbf{r}_2)\rangle. \tag{12.7.11}$$

Taking Fourier transforms

$$\Delta\eta(\mathbf{r}) = \sum_{\mathbf{k}} \Delta\eta_{\mathbf{k}} e^{i\mathbf{k}\cdot\mathbf{r}}, \tag{12.7.12}$$

$$\nabla(\Delta\eta(\mathbf{r})) = i \sum_{\mathbf{k}} \mathbf{k}\Delta\eta_{\mathbf{k}} e^{i\mathbf{k}\cdot\mathbf{r}}, \tag{12.7.13}$$

$$\langle |\Delta\eta_{\mathbf{k}}|^2 \rangle = \frac{T_c}{2V} \left(gk^2 + at\right)^{-1}, \tag{12.7.14}$$

we obtain for the correlation function

$$C(\mathbf{r}) = \frac{V}{(2\pi)^3} \int \langle |\Delta\eta_{\mathbf{k}}|^2 \rangle e^{i\mathbf{k}\cdot\mathbf{r}} d\mathbf{k} \tag{12.7.15}$$

$$= \frac{T_c}{8\pi g} \frac{e^{-\frac{r}{\xi}}}{r}, \text{ with} \tag{12.7.16}$$

$$\xi = \sqrt{\frac{g}{a|t|}}. \tag{12.7.17}$$

The **correlation radius** ξ again follows a power-law behaviour

$$\xi \propto |t|^{-\nu} \tag{12.7.18}$$

with the Landau value of this new critical exponent $\nu = \frac{1}{2}$. Fluctuations situated at distance larger than this correlation radius ξ are *not* correlated while those at separation less than ξ are correlated with each other. There is an identity for this critical index ν for a system with space dimension d

$$\nu \cdot d = 2 - \alpha. \tag{12.7.19}$$

Identity 12.7.19 is *not* satisfied by Landau theory for space dimensions $d \leq 3$. This shows that Landau theory will *not* work in the **fluctuation domain** when fluctuations are highly correlated. Space dimension $d = 4$, when Identity 12.7.19 is satisfied, has an important role in modern theory of phase transition. In this case Landau theory can be used even in the fluctuation dominated region.

We shall now find an estimate of the fluctuation domain about the critical point T_c outside of which we can apply Landau theory. The *condition of applicability* of Landau theory is

$$\langle \eta^2 \rangle \gg \langle |\Delta\eta|^2 \rangle. \tag{12.7.20}$$

We recall that the general theory of fluctuation (§1.6) gives $\langle |\Delta\eta|^2 \rangle = \frac{\chi k_B T_c}{V_c}$ where the correlation volume is $V_c = \xi^3$. In the Landau theory

$$\langle \eta^2 \rangle = \frac{a|t|}{2B} \quad \text{and} \tag{12.7.21}$$

$$\chi = \frac{V_c}{4a|t|}. \tag{12.7.22}$$

These, together with the other condition that $|t| \ll T_c$ allows us to arrive at the temperature range

$$T_c \gg |t| \gg \sqrt{\frac{k_B T_c B}{2a^2}}. \tag{12.7.23}$$

Modern theories of phase transition using Renormalization Group Technique investigates this region of large fluctuations of order parameter. The dynamics of the system is studied through the hamiltonian of the system.

12.8 Ising Model

In the study of phase transition Ising Model has perhaps the most important position. In the *one* and the *two dimensional* cases this model can be exactly solved. Attempt to get a solution of the *three dimensional* case has given rise to modern theories of phase transition. Originally this model was used to analyze magnetic systems. In the **Heisenberg theory**, the hamiltonian for a magnetic system when there is no external magnetic field is given by

$$\mathcal{H} = -2 \sum_{\langle i,j \rangle} J_{ij} \mathbf{S}_i \cdot \mathbf{S}_j. \tag{12.8.1}$$

Here $\mathbf{S}_i$ is the Spin at the i-th site, J_{ij} is the exchange integral for the sites i and j, and the summation is over the pair $\langle i, j \rangle$. For ferromagnetic interaction J is positive and for antiferromagnetic interaction J is negative. In the **Ising model** the expression within the summation is truncated to keep only the *z-component*.

$$\mathcal{H} = -2 \sum_{\langle i,j \rangle} J_{ij} S_i^z S_j^z. \tag{12.8.2}$$

If an external magnetic field $\mathbf{B}$ along the z-direction is present, then we have the modified **Ising hamiltonian**

$$\mathcal{H} = -2 \sum_{\langle i,j \rangle} J_{ij} S_i^z S_j^z - g\mu_B B \sum_i S_i^z. \tag{12.8.3}$$

Here g is the Landé g-factor and μ_B is the *Bohr magneton*. A further simplified approximation is often made by restricting the inter-spin interaction only to *nearest-neighbors*: $J_{ij} = J$ only when the sites i and j are nearest neighbors, and zero otherwise.

This hamiltonian is used not only for studying the magnetic systems, but various other physical and even biophysical systems. For example, in the **Lattice gas model** of *random binary alloys* with 2 components A & B and using the notations:

p_i = probability that the i − th site is occupied with values 0 or 1,

$2\epsilon_{ij}$ = interaction energy between atoms at the i − th & the j − th sites.,

we get the expression of energy

$$E_p = -2 \sum_{\langle i,j \rangle} [\epsilon_{ij}\,(AA)\,p_i\,(A)\,p_j\,(A) + \epsilon_{ij}\,(AB)\,(p_i\,(A)\,p_j\,(B) + p_i\,(B)\,p_j\,(A))$$

$$+ \epsilon_{ij}\,(BB)\,p_i\,(B)\,p_j\,(B)]$$

$$= -2 \sum_{\langle i,j \rangle} [2\epsilon_{ij}\,(AB) - \epsilon_{ij}\,(AA) - \epsilon_{ij}\,(BB)]\,p_i\,(A)\,p_j\,(B)$$

+ terms independent of configurations of A and B atoms.   (12.8.4)

In arriving at Equation 12.8.4 we have used

$$p_i\,(A) + p_i\,(B) = 1, \qquad \text{for all i.} \tag{12.8.5}$$

$$\tag{12.8.6}$$

When we compare Equations 12.8.2 and 12.8.4 we immediately notice their essential similarity.

In the zero-field ferromagnetic case with spin-1/2 case, when

$$J > 0,\ S_i^z = \pm\frac{1}{2},\ B = 0 \tag{12.8.7}$$

the *Partition function* becomes

$$\mathcal{Z} = \sum_{\sigma_1 = \pm 1} \sum_{\sigma_2 = \pm 1} \cdots \prod_{\langle i,j \rangle} \exp\,(L\sigma_i\sigma_j)$$

$$= \sum_{\sigma_1 = \pm 1} \sum_{\sigma_2 = \pm 1} \cdots \prod_{\langle i,j \rangle} [\cosh\,(L) + \sigma_i\sigma_j \sinh\,(L)], \tag{12.8.8}$$

where

$$L = \frac{J}{2k_BT} \tag{12.8.9}$$

Equation 12.8.8 results from

$$\exp\left(L\sigma_i\sigma_j\right) = \cosh\left(L\right) + \sigma_i\sigma_j\left(L\right),$$

since

$$\sigma_i\sigma_j = \pm 1.$$

12.8.1 Zero-Field 1-Dimensional Case

We first calculate the 1-dimensional case with N atoms depicted in Figure 12.8.1. In calculating $\mathcal{Z}$ we note that $\sum_{\sigma_i=\pm 1}\sigma_i = 0$ and $\sigma_i^2 = 1$ for all

$$\sigma_1 \qquad \sigma_2 \qquad \sigma_3 \qquad\qquad\qquad \sigma_{N-2} \qquad \sigma_{N-1} \qquad \sigma_N$$

Fig. 12.8.1   1-dimensional Ising chain with N atoms.

i. Thus

$$\mathcal{Z} = [\cosh(L)]^N \sum_{\sigma_1=\pm 1} \cdots \sum_{\sigma_N=\pm 1} \prod_{i=1}^{N}[1 + \sigma_i\sigma_{i+1}\tanh(L)]. \qquad (12.8.10)$$

When both ends of the chain are open we have $\sigma_{N+1} = 0$ and the summation over $\sigma_i = \pm 1$ of the products of any number terms of the form $\sigma_i\sigma_{i+1}$ is zero; hence $\mathcal{Z} = [2\cosh(L)]^N$. For periodic boundary condition $\sigma_{N+1} = \sigma_1$ and only the term $\sigma_1\sigma_2\sigma_2\sigma_3\sigma_3\cdots\sigma_N\sigma_N\sigma_1 = 1$ is non-vanishing. In this case $\mathcal{Z} = [2\cosh(L)]^N + [2\sinh(L)]^N$. We get in both the cases for finite L

$$\lim_{N\to\infty}\mathcal{Z} = [2\cosh(L)]^N \qquad (12.8.11)$$

is a smooth function of L and hence of temperature T. Thus no phase transition occurs at *finite temperature*.

12.8.2 Non-Zero-Field 1-Dimensional Case

The 1-dimensional ising model in the presence of external magnetic field

$$\mathcal{H} = -\frac{g\mu_B B}{2}\sum_{i=1}^{N}\sigma_i - \frac{J}{2}\sum_{i=1}^{N}N\sigma_i\sigma_{i+1}, \text{with } \sigma_i = \pm 1 \qquad (12.8.12)$$

can also be exactly solved. The calculation by *matrix method* becomes easy if we work with cyclic boundary condition, $\sigma_{N+1} = \sigma_1$. The canonical partion function can be written as

$$\mathcal{Z} = \sum_{\sigma_1 = \pm 1} \cdots \sum_{\sigma_N = \pm 1} K\left(\sigma_1, \sigma_2\right) K\left(\sigma_2, \sigma_3\right) \cdots K\left(\sigma_{N-1}, \sigma_N\right) K\left(\sigma_N, \sigma_1\right),$$

$$(12.8.13)$$

where

$$K\left(\sigma_i, \sigma_{i+1}\right) = \exp\left[\frac{C}{2}\left(\sigma_i + \sigma_{i+1}\right) + L\sigma_i\sigma_{i+1}\right] \qquad (12.8.14)$$

and

$$C = \frac{g\mu_B B}{2k_B T}, \quad L = \frac{J}{2k_B T}. \qquad (12.8.15)$$

If we use the *real symmetric matrix*

$$\mathbf{K} = \begin{pmatrix} \exp\left(C + L\right) & \exp\left(-L\right) \\ \exp\left(-L\right) & \exp\left(-C + L\right) \end{pmatrix} \qquad (12.8.16)$$

with eigenvalues

$$\lambda_\pm = \exp\left(L\right)\cosh\left(C\right) \pm \sqrt{\exp\left(2L\right)\sinh^2\left(C\right) + \exp\left(-2L\right)}, \qquad (12.8.17)$$

where

$$|\lambda_+| > |\lambda_-|, \qquad (12.8.18)$$

then in the limit of infinite chain the partition function reduces to

$$\begin{aligned}
\mathcal{Z}_\infty &= \lim_{N \to \infty} \mathcal{Z} \\
&= \lim_{N \to \infty} Tr\left(\mathbf{K}^N\right) \\
&= \lim_{N \to \infty} \left(\lambda_+^N + \lambda_-^N\right) = \lambda_+^N \\
&= \left[\exp\left(L\right)\cosh\left(C\right) + \sqrt{\exp\left(2L\right)\sinh^2\left(C\right) + \exp\left(-2L\right)}\right]^N.
\end{aligned}$$

$$(12.8.19)$$

The magnetization turns out to be

$$M = -\frac{\partial}{\partial B}\left(-k_B T \ln \mathcal{Z}_\infty\right) = \frac{N g\mu_B}{2} \frac{\sinh\left(C\right)}{\sqrt{\exp\left(-4L\right) + \sinh^2\left(C\right)}}. \qquad (12.8.20)$$

As the external magnetic field B tends to zero (i.e. $C \to 0$) M tends to zero showing absence of an ordered state with *spontaneous magnetization*. However, a paramagnetic susceptibility

$$\chi = \left(\frac{\partial M}{\partial B}\right)_{B \to 0} = \frac{N (g\mu_B)^2}{4k_B T} \exp\left(\frac{J}{k_B T}\right) \tag{12.8.21}$$

exists. For high temperature $k_B T \gg |J|$, this susceptibility satisfies **Curie's law**.

12.8.3 *Multi-Dimensional Case*

We now discuss Ising model in multi-dimension. Since a thorough discussion requires complex topological analysis we shall restrict us to a general qualitative discussion and shall simply point out the results without any rigorous mathematical proof.

In 1-dimension any atom has only 2 nearest neighbours. Let, in general,

$$c = \text{number of nearest neighbours of a lattice point.}$$

If the total number of lattice points is N, then

$$s = \text{total number of nearest neighbour pairs} = cN/2,$$

and Equation 12.8.8 reduces to

$$
\begin{aligned}
\mathcal{Z} &= (\cosh{(L)})^s \sum_{\sigma_1 = \pm 1} \cdots \sum_{\sigma_N = \pm 1} \prod_{\langle i,j \rangle} [1 + \sigma_i \sigma_j \tanh{(L)}] \\
&= (\cosh{(L)})^s \sum_{\sigma_1 = \pm 1} \cdots \sum_{\sigma_N = \pm 1} \left[1 + (\tanh{(L)}) \sum_{\langle i_1, j_1 \rangle} (\sigma_{i_1} \sigma_{j_1}) \right. \\
&\quad + (\tanh{(L)})^2 \sum_{\langle i_1, j_1 \rangle} \sum_{\langle i_2, j_2 \rangle} (\sigma_{i_1} \sigma_{j_1})(\sigma_{i_2} \sigma_{j_2}) + \cdots \\
&\quad \left. + (\tanh{(L)})^s \sum_{\langle i_1, j_1 \rangle} \cdots \sum_{\langle i_s, j_s \rangle} (\sigma_{i_1} \sigma_{j_1}) \cdots (\sigma_{i_s} \sigma_{j_s}) \right]
\end{aligned} \tag{12.8.22}
$$

In the sums over different pairs on the right hand side of Equation 12.8.22 no pair appears more than once. Moreover if some σ appears an odd number of times then the sum over that particular σ makes the term vanish. Thus non-zero contribution to the partition function $\mathcal{Z}$ comes only from pairs forming closed polygons. So we can write the partition function as

$$\mathcal{Z} = 2^N \left(\cosh\left(L\right)\right)^s \left[1 + \sum_n \Omega_n \left(\tanh\left(L\right)\right)^n\right] \qquad (12.8.23)$$

where

$$\Omega_n = \text{number of closed polygon with n bonds.} \qquad (12.8.24)$$

Because of the nature of the terms on the right hand side of Equation 12.8.23 we see that it is a *high temperature expansion* of the partition function. In 1-dimension $\Omega_n = 0$ for all $n > 0$ except for Ω_N under periodic boundary condition. In 2-dimensional square lattice and 3-dimensional simple cubic lattice only terms with even n remain in the expansion.

A *low temperature expansion* can be arrived at from an alternate consideration. We assign an up spin inside a closed polygon of n bonds, so that these up spins form the dual lattice; and put down spins outside the polygons, so that connecting pairs of antiparallel spins on the dual lattice intersect the n bonds of the original direct lattice. We now define

$$r = \text{number of antiparallel spin pairs } \langle\uparrow\downarrow\rangle, \qquad (12.8.25)$$

$$s - r = \text{number of parallel spin pairs } \langle\uparrow\uparrow\rangle \qquad (12.8.26)$$

so that

$$\sum_{\langle i,j\rangle} \sigma_i\sigma_j = (s - r) - r = s - 2r \qquad (12.8.27)$$

and

$$\mathcal{Z} = 2e^{sL} \left[1 + \sum_r \omega_r e^{-2rL}\right], \qquad (12.8.28)$$

where

$$\omega_r = \text{number of configurations with } r \text{ antiparallel spin pairs}$$

For ferromagnetic systems when $J > 0$, e^{-2rL} becomes small at low temperatures.

12.8.4   2-Dimensional Ising System

Ising system in 2-dimension has a special position in the theory of phase transition, ever since Onsager pointed out and Yang analytically proved that this system has a phase transition at finite temperature.

We first show in Figure 12.8.2 typical examples of 2-dimensional polygons that have zero or non-zero contributions to the partition function $\mathcal{Z}$.

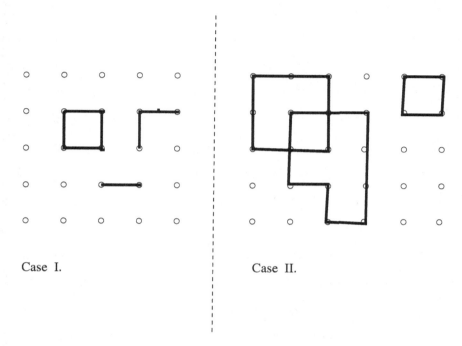

Fig. 12.8.2 Typical examples of closed and open polygons in a 2-dimensional square lattice. Case I: open polygons that have zero contribution to Z. Case II: closed polygons that have non-zero contribution to Z.

In Figure 12.8.3 we now show two examples of direct lattices in two dimension and their dual lattices. For the dual lattice the notation for a particular quantity is denoted by putting an asterisk to the notation for the same quantity in the direct lattice. Thus

$$\Omega_n = \omega_n^*, \quad \omega_n = \Omega_n^*, \text{ and of course } s = s^*. \qquad (12.8.29)$$

With these conventions Equations 12.8.23 and 12.8.28 become

$$\mathcal{Z}(T) = 2^N \left(\cosh\left(L\right)\right)^s \left[1 + \sum_n \Omega_n \left(\tanh\left(L\right)\right)^n\right], \qquad (12.8.30)$$

$$\mathcal{Z}^*\left(T\right) = 2e^{sL}\left[1 + \sum_n \Omega_n e^{-2nL}\right]. \qquad (12.8.31)$$

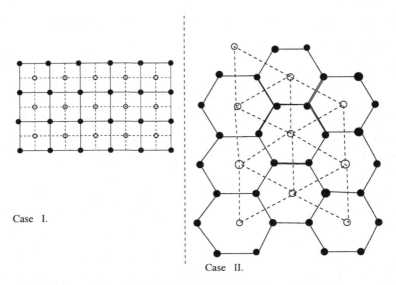

Fig. 12.8.3 Two examples of direct lattices in 2-dimension. Case I: A square lattice with bonds depicted by solid lines and its dual lattice (also a square lattice) with bonds depicted by broken lines. Case II: A honeycomb lattice with bonds depicted by solid lines and its dual lattice (a triangular lattice) with bonds depicted by broken lines.

If we define L^* and T^* by

$$\tanh(L) = e^{-2L^*}, \quad \text{and} \quad T^* = \frac{J}{2k_B L^*} \tag{12.8.32}$$

and use the *symmetry relations*

$$\tanh(L^*) = e^{-2L}, \tag{12.8.33}$$

$$\sinh(2L)\sinh(2L^*) = 1, \tag{12.8.34}$$

$$\tanh(2L^*)\cosh(2L) = \tanh(2L)\cosh(2L^*) = 1, \tag{12.8.35}$$

we get

$$\mathcal{Z}(T) = 2^{N-1}\left(\cosh(L)\right)^s e^{-sL^*} \mathcal{Z}^*(T^*) i, \tag{12.8.36}$$

$$\frac{\mathcal{Z}(T)}{2^{(N-1)/2}\left(\cosh(2L)\right)^{\frac{s}{2}}} = \frac{\mathcal{Z}^*(T^*)}{2^{(N^*+1)/2}\left(\cosh(2L^*)\right)^{\frac{s}{2}}}. \tag{12.8.37}$$

In arriving at the last equation we have to use *Euler's Theorem of polyhedra*:

$$N + N^* = s.$$

We now mention Onsager's results of critical points of an Ising system in 2-dimensions. On increasing temperature T (which corresponds to lowering of T^*) and approaching the critical point T_c where $\mathcal{Z}(T)$ has a singularity, $\mathcal{Z}^*(T^*)$ approaches its singularity at $T^* = T_c^*$. Detailed analysis shows that for

- Square Lattice: $\sinh^2(2L_c) = 1$, leading to $L_c = \frac{J}{2k_B T_c} = 0.4407\cdots$,
- Triangular Lattice: $\exp(4L_c) = 3$, leading to $L_c = 0.2747\cdots$,
- Honeycomb Lattice: $\exp(2L_c) = 2$, leading to $L_c = 0.6585\cdots$.

12.9 Problems

Problem 12.1. Prove Equations 12.7.14 and 12.7.16.

Problem 12.2. Prove Equation 12.7.23.

Problem 12.3. Prove Equation 12.8.8.

Problem 12.4. Prove Equation 12.8.13.

Problem 12.5. Prove Equation 12.8.20.

Problem 12.6. Prove Equation 12.8.21.

Problem 12.7. Prove Equation 12.8.22.

Chapter 13

Irreversible Processes

13.1 Introduction

In Kinetic Theory of Gases we study in detail two prototypes of **irreversible processes**: Thermal Conduction and Diffusion. In Classical Electrodynamics Electrical Conduction is another important example of *irreversible process*. In each of these phenomena there is an appropriate *force field* (temperature gradient $-\nabla T$, density gradient $-\nabla n$ and electrical field $\mathbf{E}$) that maintains a constant *current* (thermal current $\mathbf{j}_{\text{thermal}}$, particle current $\mathbf{j_n} = n\mathbf{u}$ and electric charge current $\mathbf{j}_{\text{el}} = \rho_{\text{el}}\mathbf{u}$) and a **Kinetic Coefficient** (thermal conductivity tensor $\tilde{K}$, diffusion coefficient D and electrical conductivity tensor $\tilde{\sigma}$) that connects the current to the force field. Similarly, in magnetism magnetic susceptibility tensor $\tilde{\chi}$ connects the magnetization $\mathbf{M}$ to the applied magnetic induction $\mathbf{B}$.

$$\mathbf{j}_{\text{thermal}} = -\tilde{K}\nabla T \quad \text{for thermal conduction,} \qquad (13.1.1)$$

$$\mathbf{j_n} = -D\nabla n \quad \text{Fick's law for diffusion,} \qquad (13.1.2)$$

$$\mathbf{j}_{\text{el}} = \tilde{\sigma}\mathbf{E} \quad \text{for electrical conduction,} \qquad (13.1.3)$$

$$\mathbf{M} = \tilde{\chi}\mathbf{B} \quad \text{for magnetic response.} \qquad (13.1.4)$$

The well-known equation for diffusion follows from Equation 13.1.2 when we use the equation of continuity

$$\frac{\partial n}{\partial t} + \nabla \cdot \mathbf{j_n} = 0. \qquad (13.1.5)$$

Statistical Physics generalizes these concepts to **Generalized Force, Generalized Coordinate** and **Generalized Susceptibility/Admittance**.

In the traditional method we calculate the Kinetic Coefficients by using the Kinetic equation, which itself is an approximation and is *not* always derivable without some stringent and often unsatisfactory restrictions. In the **Linear Response Theory** of Statistical Physics, Generalized Admittance is calculated by solving the Liouville's equation and calculating equilibrium thermal average of an appropriate fluctuation. This is similar to Heisenberg-Kramers dispersion formula for light scattering.

13.2 Linear Response Theory (Kubo Formalism)

Irreversible processes are classified in two types:

(i) *Mechanical Process*, when the external generalized force is described by a *hamiltonian*.

(ii) *Thermal Process*, when the external generalized force is *not* described by a hamiltonian. Thermal conduction and diffusion belong to this class.

13.2.1 *Mechanical Process*

In the **mechanical process** the **natural motion** of the system is described by the *hamiltonian* $\hat{\mathcal{H}}_0$ and the external generalized force is described by the perturbation (assumed to be weak),

$$\hat{\mathcal{H}}'(t) \; = \; -\hat{A}F(t). \tag{13.2.1}$$

We are calculating the change $\Delta\hat{B}(t)$ of the observable $\hat{B}(t)$ as a result of the perturbation. The unperturbed density matrix $\hat{\rho}_0$ satisfies

$$\left[\hat{\mathcal{H}}_0, \hat{\rho}_0\right] \; = \; 0. \tag{13.2.2}$$

The non-equilibrium value of the density matrix $\hat{\rho}$ satisfies

$$\hat{\rho}(t) = \hat{\rho}_0 \; + \; \Delta\hat{\rho}(t), \tag{13.2.3}$$

$$i\hbar\frac{\partial\hat{\rho}(t)}{\partial t} \; = \; \left[\hat{\mathcal{H}}_0 \; + \; \hat{\mathcal{H}}'(t), \hat{\rho}(t)\right] \tag{13.2.4}$$

with boundary condition

$$\hat{\rho}(-\infty) \; = \; \hat{\rho}_0. \tag{13.2.5}$$

In the **interaction (Dirac) picture**

$$\hat{\rho}_D(t) = e^{\frac{+i\hat{\mathcal{H}}_0 t}{\hbar}} \hat{\rho}(t) e^{\frac{-i\hat{\mathcal{H}}_0 t}{\hbar}}, \tag{13.2.6}$$

$$\Delta\hat{\rho}_D(t) = e^{\frac{+i\hat{\mathcal{H}}_0 t}{\hbar}} \Delta\hat{\rho}(t) e^{\frac{-i\hat{\mathcal{H}}_0 t}{\hbar}}, \tag{13.2.7}$$

$$\hat{\mathcal{H}}'_D(t) = e^{\frac{+i\hat{\mathcal{H}}_0 t}{\hbar}} \hat{\mathcal{H}}'(t) e^{\frac{-i\hat{\mathcal{H}}_0 t}{\hbar}}. \tag{13.2.8}$$

In the *linear approximation*, we get the equation of motion for $\Delta\hat{\rho}_D(t)$ and hence the expression for $\Delta\hat{\rho}(t)$

$$i\hbar\frac{\partial\Delta\hat{\rho}_D(t)}{\partial t} = \left[\hat{\mathcal{H}}'_D(t), \hat{\rho}_0\right], \tag{13.2.9}$$

$$\Delta\hat{\rho}(t) = (i\hbar)^{-1} \int_{-\infty}^{t} dt' e^{\frac{-i\hat{\mathcal{H}}_0(t-t')}{\hbar}} \left[\hat{\mathcal{H}}'(t'), \hat{\rho}_0\right] e^{\frac{+i\hat{\mathcal{H}}_0(t-t')}{\hbar}}$$

$$= -(i\hbar)^{-1} \int_{-\infty}^{t} dt' e^{\frac{-i\hat{\mathcal{H}}_0(t-t')}{\hbar}} \left[\hat{A}, \hat{\rho}_0\right] e^{\frac{+i\hat{\mathcal{H}}_0(t-t')}{\hbar}} F(t'). \tag{13.2.10}$$

For the response $\Delta\hat{B}(t)$ to the observable $\hat{B}$ we get

$$\langle\Delta\hat{B}(t)\rangle = Tr\left(\Delta\hat{\rho}(t)\hat{B}\right) \tag{13.2.11}$$

$$= -(i\hbar)^{-1} Tr\left(\int_{-\infty}^{t} dt' \left[\hat{A}, \hat{\rho}_0\right] \hat{B}_D(t-t') F(t')\right). \tag{13.2.12}$$

In terms of the **Response Function**

$$\phi_{\hat{B},\hat{A}}(t) \overset{\text{def}}{=} -(i\hbar)^{-1} Tr\left(\left[\hat{A}, \hat{\rho}_0\right] \hat{B}_D(t)\right) \tag{13.2.13}$$

$$= (i\hbar)^{-1} Tr\left(\hat{\rho}_0\left[\hat{A}, \hat{B}_D(t)\right]\right). \tag{13.2.14}$$

Equation 13.2.12 can be expressed as

$$\langle\Delta\hat{B}(t)\rangle = \int_{-\infty}^{t} dt' \, \phi_{\hat{B},\hat{A}}(t-t') F(t'). \tag{13.2.15}$$

In classical cases there is no difficulty in interpreting the response as the average closely approximating the observed value. In quantum mechanical case some care has to be taken. We prepare an ensemble of subsystems and measure $\hat{B}$ for some of the subsystems at time t; at time t' we choose some other subsystems and measure the change $\Delta\hat{B}$. The first set of subsystems are *not* followed after measurents have been performed on them.

We thus observe time evolution of $\Delta \hat{B}(t)$. For macroscopic bodies quantum mechanical disturbance may be neglected; but exact conditions when this can be done is *not* known.

For a simple periodic perturbation

$$F(t) = F_0 \cos \omega_0 t = \Re\left(F_0 e^{i\omega_0 t}\right). \tag{13.2.16}$$

Equation 13.2.15 can be written as

$$\langle \Delta \hat{B}(t) \rangle = \Re\left(\chi_{\hat{B},\hat{A}}(\omega_0) F_0 e^{i\omega_0 t}\right), \tag{13.2.17}$$

where we have defined the **Generalized Susceptibility/Admittance** as

$$\chi_{\hat{B},\hat{A}}(\omega) \stackrel{\text{def}}{=} \lim_{\epsilon \to 0^+} \int_0^\infty \phi_{\hat{B},\hat{A}}(t) e^{-i\omega t - \epsilon t} dt. \tag{13.2.18}$$

In general when there are many frequancy components in the spectrum of the external perturbation

$$F(t) = \sum_\Omega F_\Omega e^{i\Omega t}, \tag{13.2.19}$$

the response becomes

$$\langle \Delta \hat{B}(t) \rangle = \sum_\Omega \chi_{\hat{B},\hat{A}}(\Omega) F_\Omega e^{i\Omega t}. \tag{13.2.20}$$

When $\lim_{t\to\infty} \phi_{\hat{B}\hat{A}}(t)$ is finite we can use **Abel's Theorem**

$$\lim_{t\to\infty} \phi_{\hat{B}\hat{A}}(t) = \lim_{\epsilon\to 0^+} \epsilon \int_0^\infty \phi_{\hat{B}\hat{A}}(t) e^{-\epsilon t} dt \tag{13.2.21}$$

to obtain

$$\lim_{t\to\infty} \phi_{\hat{B}\hat{A}}(t) = \phi_{\hat{B}\hat{A}}(\Omega = 0), \tag{13.2.22}$$

where the *Fourier Transform* $\phi_{\hat{B}\hat{A}}(\Omega)$ is defined by

$$\phi_{\hat{B}\hat{A}}(t) = \sum_\Omega \phi_{\hat{B}\hat{A}}(\Omega) e^{i\Omega t}. \tag{13.2.23}$$

Elimination of long **Poincaré Cycles** is thus equivalent to having the limit

$$\lim_{t\to\infty} \phi_{\hat{B}\hat{A}}(t) = \phi_{\hat{B}\hat{A}}(\Omega = 0) = 0. \tag{13.2.24}$$

In many experimental set-ups a constant external perturbation F acts from $t = -\infty$ up to $t = 0$ and is then cut off. The response $\Delta B(t)$ will then relax according to the formula

$$\langle \Delta \hat{B}(t) \rangle = \int_{-\infty}^{0} dt' \, \phi_{\hat{B}, \hat{A}}(t - t') \, F(t')$$

$$= F \int_{t}^{\infty} \phi_{\hat{B}\hat{A}}(t') \, dt' \quad t \geq 0 \tag{13.2.25}$$

$$= F \Phi_{\hat{B}\hat{A}}(t), \tag{13.2.26}$$

where the **Relaxation Function**

$$\Phi_{\hat{B}\hat{A}}(t) \stackrel{\text{def}}{=} \lim_{\epsilon \to 0^+} \int_{t}^{\infty} \phi_{\hat{B}\hat{A}}(t') \, e^{-\epsilon t'} \, dt' \tag{13.2.27}$$

describes the evolution of $\langle \Delta \hat{B}(t) \rangle$ after the external perturbation ceases to work. In terms of the **Relaxation Function**, the expression for the **Generalized Susceptibility** can be written as

$$\chi_{\hat{B}\hat{A}}(\omega) = \Phi_{\hat{B}\hat{A}}(t = 0) - i\omega \int_{0}^{\infty} \Phi_{\hat{B}\hat{A}}(t) \, e^{-i\omega t} dt. \tag{13.2.28}$$

In case of a *canonical ensemble*, which is the most prevalent experimental situation, the results are obtained in the following way. First we use Theorem 14.4.1 of § 14.4, the Mathematical Appendix, to write

$$e^{-\beta \hat{\mathcal{H}}_0} \int_{0}^{\beta} e^{+\lambda \hat{\mathcal{H}}_0} \left[\hat{\mathcal{H}}_0, \hat{A} \right] e^{-\lambda \hat{\mathcal{H}}_0} d\lambda$$

$$= e^{-\beta \hat{\mathcal{H}}_0} \sum_{m=0}^{\infty} \frac{1}{m!} \int_{0}^{\beta} \lambda^m d\lambda \left[\hat{\mathcal{H}}_0, \left[\hat{\mathcal{H}}_0, \cdots, \left[\hat{\mathcal{H}}_0, \left[\hat{\mathcal{H}}_0, \hat{A} \right] \right] \cdots \right] \right]_{m+1 \text{ terms}}$$

$$= e^{-\beta \hat{\mathcal{H}}_0} \left(e^{+\beta \hat{\mathcal{H}}_0} \hat{A} e^{-\beta \hat{\mathcal{H}}_0} - \hat{A} \right) = \left[\hat{A}, e^{-\beta \hat{\mathcal{H}}_0} \right]. \tag{13.2.29}$$

Using Equation 14.8.12 and Equation 14.8.13 of § 14.8, we get the relation

$$\left[e^{-\beta \mathcal{H}_0}, \hat{A} \right] = (i\hbar) \int_{0}^{\beta} d\lambda \dot{\hat{A}}(-i\hbar\lambda). \tag{13.2.30}$$

Remembering that for canonical ensembles $\hat{\rho}_0 = e^{-\beta \hat{\mathcal{H}}_0} / Tr\left(e^{-\beta \hat{\mathcal{H}}_0}\right)$ where $\beta = 1/k_B T$, T being absolute temperature, Equation 13.2.14 for the **Response Function** reduces to the **Kubo formula**

$$\phi_{\hat{B}\hat{A}}(t) = (-i\hbar)^{-1} Tr\left[\hat{A}, \hat{\rho}_0\right]\hat{B}_D(t)$$

$$= \int_0^\beta Tr\left(\hat{\rho}_0\hat{A}_D(-i\hbar\lambda)\hat{B}_D(t)\right)d\lambda$$

$$= -\int_0^\beta Tr\left(\hat{\rho}_0\hat{A}_D(-i\hbar\lambda)\dot{\hat{B}}_D(t)\right)d\lambda. \qquad (13.2.31)$$

For the *Relaxation Function* we have the corresponding *Kubo Formula*

$$\Phi_{\hat{B}\hat{A}}(t) = -\int_0^\beta d\lambda Tr\left(\hat{\rho}_0\hat{A}_D(-i\hbar\lambda)\int_t^\infty dt'\dot{\hat{B}}_D(t')\right)$$

$$= \int_0^\beta d\lambda Tr\left(\hat{\rho}_0\hat{A}_D(-i\hbar\lambda)\hat{B}_D(t)\right)$$

$$- \int_0^\beta d\lambda \lim_{t\to\infty} Tr\left(\hat{\rho}_0\hat{A}_D(-i\hbar\lambda)\hat{B}_D(t)\right)$$

$$= \int_0^\beta d\lambda Tr\left(\hat{\rho}_0\hat{A}_D(-i\hbar\lambda)\hat{B}_D(t)\right) - \beta Tr\left(\hat{\rho}_0\hat{A}^d\hat{B}^d\right).$$

$$(13.2.32)$$

Here $\hat{A}^d$ and $\hat{B}^d$ are the diagonal parts of the corresponding operators in the energy-representation.

From Equation 13.2.28 we obtain the **static** value of the *Generalized Susceptibility/Admittance*

$$\chi_{\hat{B}\hat{A}}(\omega = 0) = \Phi_{\hat{B}\hat{A}}(t = 0)$$

$$= \int_0^\beta d\lambda Tr\left(\hat{\rho}_0\hat{A}_D(-i\hbar\lambda)\hat{B}\right) - \beta Tr\left(\hat{\rho}_0\hat{A}^d\hat{B}^d\right)$$

$$= \int_0^\beta d\lambda Tr\left(\hat{\rho}_0\left(\hat{A}_D(-i\hbar\lambda) - \hat{A}^d\right)\left(\hat{B} - \hat{B}^d\right)\right).$$

$$(13.2.33)$$

It is to be noted that this static admittance given in Equation 13.2.33 is *not in general* equal to the **isothermal** value of the General Susceptibility/Admittance :

$$\chi_{\hat{B}\hat{A}}^T(\omega = 0) = \int_0^\beta d\lambda Tr\left(\hat{\rho}_0\left(\hat{A}_D(-i\hbar\lambda) - \langle\hat{A}^d\rangle\right)\left(\hat{B} - \langle\hat{B}^d\rangle\right)\right).$$

$$(13.2.34)$$

Here $\langle\hat{A}^d\rangle$ and $\langle\hat{B}^d\rangle$ are the ensemble averages of the relevant operators.

Ever since the publication by Kubo of these results there have been intense discussions on them, particularly about the domain of validity of the linear approximation and the role of *stochastization* involved in the process. However, those points have more or less been satisfactorily dealt with and the Linear Response Theory and its subsequent development by Redfield is an essential tool for analysis of irreversible processes, especially by Resonance and Relaxation community.

13.2.2 *Thermal Process*

In the case of mechanical processes we have obtained expressions for kinetic coefficients in terms of ensemble averages over the unperturbed density matrix. There are other irreversible processes like *Diffusion* and *Thermal Conductivity*, called *Thermal Processes*, which cannot be described by an external hamiltonian. Similar expressions for the kinetic coefficients in these cases can also be given if we invoke the **Onsager's Postulate**.

Postulate 13.2.1 (Onsager's). The average behaviour of fluctuations of a physical quantity in an aged system is governed by the macroscopic physical law which governs the macroscopic change of the corresponding macroscopic variable.

We shall now see that this Postulate provides us with a method of calculating the kinetic coefficients in terms of the time correlation of a set of stochastic fluctuating macroscopic variables $\hat{\alpha}_j (j = 1, 2, \cdots, n)$ if the correlation can be calculated.

Furthermore, $\overline{\hat{\alpha}_j(t + \Delta t | \alpha')}$ defines the expectation value of the stochastic variable $\hat{\alpha}_j$ at $t + \Delta t$ for an ensemble when the variables $(\hat{\alpha}_1, \hat{\alpha}_2, \cdots, \hat{\alpha}_n)$ have the expectation values $(\alpha'_1, \alpha'_2, \cdots, \alpha'_n)$ at time t. This ensemble has the density matrix $\hat{\rho}(t + \Delta t | \alpha')$, satisfying

$$Tr\left(\hat{\rho}(t + 0 | \alpha')\hat{\alpha}_j\right) = \alpha'_j, \tag{13.2.35}$$

$$Tr\left(\hat{\rho}(t + \Delta t | \alpha')\hat{\alpha}_j\right) = \overline{\hat{\alpha}_j(t + \Delta t | \alpha')}. \tag{13.2.36}$$

We can now mathematically write Onsager's Postulate 13.2.1 as

$$Tr\left(\hat{\rho}(t + \Delta t | \alpha')\hat{\alpha}_j\right) - \alpha'_j = \sum_k G_{jk}\frac{\partial S}{\partial \alpha'_k}\Delta t. \tag{13.2.37}$$

Here S is the entropy and G_{jk} are the macroscopic **kinetic coefficient tensors**. The problem is the form of the density matrix $\hat{\rho}(t + \Delta t | \alpha')$. The natural choice for this density matrix

$$\hat{\rho}\left(t + \Delta t | \alpha'\right) = \exp\left[\beta\left(F - \hat{\mathcal{H}}\right) - \beta \sum_j A_j \hat{\alpha}_j\right] \qquad (13.2.38)$$

is that for a sort of *canonical ensemble* with Free Energy F and the generalized forces A_j $(j = 1, 2. \cdots . n)$ associated with the coordinates $\hat{\alpha}_j$ $(j = 1, 2, \cdots, n)$.

The density matrix, Equation 13.2.38, is the solution of the maximization of the entropy

$$S = k_B Tr\left(\hat{\rho} \ln \hat{\rho}\right). \qquad (13.2.39)$$

Noting that

$$\beta \frac{\partial F}{\partial A_j} = \overline{\alpha_j} \equiv \alpha'_j \qquad (13.2.40)$$

$$S = \frac{1}{T}\left[-F + \overline{\mathcal{H}} + \sum_j A_j \overline{\alpha_j}\right] \qquad (13.2.41)$$

and

$$A_j = T \frac{\partial S}{\partial \alpha'_j}, \qquad (13.2.42)$$

where the bars over any quantity means the average value for the ensemble represented by Equation 13.2.38, Equation 13.2.37 becomes

$$Tr\left[e^{\beta(F - \mathcal{H}) - \beta \sum_l A_l \alpha_l}\left(\alpha_j\left(\Delta t\right) - \alpha_j\left(0\right)\right)\right] = k_B \beta \sum_l G_{jl} A_l \Delta t. \qquad (13.2.43)$$

For a linear dissipative system where the deviation from equilibrium is small, *i.e.* the forces A_js are small we can expand the left-hand side of the last equation and keep only the linear terms. We now use the formula

$$e^{-\beta\left(\mathcal{H} + \sum_l A_l \alpha_l\right)} = e^{-\beta\mathcal{H}} - \int_0^\beta e^{-\lambda\mathcal{H}} \sum_l A_l \alpha_l\left(-i\hbar\lambda\right) \, d\lambda \qquad (13.2.44)$$

and obtain

$$\overline{\alpha_j\left(t + \Delta t | \alpha'\right)} = \langle \alpha_j \rangle - \sum_l A_l \int_0^\beta \langle \alpha_l\left(-i\hbar\lambda\right) \alpha_j\left(\Delta t\right)\rangle \, d\lambda. \qquad (13.2.45)$$

Here $\langle \alpha_j \rangle$ is the equilibrium value of α_j. At $\Delta t = 0$ we now have

$$\overline{\alpha_j\left(t | \alpha'\right)} = \alpha'_j = \langle \alpha_j \rangle - \sum_l A_l \int_0^\beta \langle \alpha_l\left(-i\hbar\lambda\right) \alpha_j \rangle \, d\lambda. \qquad (13.2.46)$$

From these equations we can conclude

$$\frac{1}{\beta} \int_0^\beta \langle \alpha_l \left(-i\hbar\lambda\right) \left(\alpha_j \left(\Delta t\right) - \alpha_j \left(0\right)\right)\rangle \, d\lambda \;=\; -k_B G_{jl} \Delta t. \qquad (13.2.47)$$

After some more mathematics it is possible to arrive at the expression for the *kinetic coefficient* in case of a thermal irreversible process

$$G_{jl} \;=\; (k_B \beta)^{-1} \int_0^{\Delta t} \left(1 - \frac{\tau}{\Delta t}\right) \, d\tau \int_0^\beta \langle \dot{\alpha}_l \left(-i\hbar\lambda\right) \dot{\alpha}_j \left(\tau\right)\rangle \, d\lambda. \qquad (13.2.48)$$

Equation 13.2.48 is an exact relation. If moreover the characteristic time of decay of the correlation functions in the integrand on the right hand side of Equation 13.2.48 is small compared to that of $\langle \alpha_j \alpha_l \rangle$ the Equation 13.2.48 reduces to

$$G_{jl} \;=\; (k_B \beta)^{-1} \int_0^\infty d\tau \int_0^\beta \langle \dot{\alpha}_l \left(-i\hbar\lambda\right) \dot{\alpha}_j \left(\tau\right)\rangle \, d\lambda. \qquad (13.2.49)$$

13.3 Symmetry Relations

The Relaxation functions and the Admittance have some useful **symmetry relations**. The reality of physical variables and *stationarity* of the correlation functions are at the base of these symmetry relations.

Theorem 13.3.1 *The Relaxation Function* $\Phi_{\hat{B}\hat{A}}(t)$ *has the following properties.*

(i)

$$\Phi_{\hat{B}\hat{A}}(t) \;=\; \Phi^*_{\hat{B}\hat{A}(t)}. \qquad (13.3.1)$$

(ii)

$$\Phi_{\hat{B}\hat{A}}(-t) \;=\; \Phi_{\hat{A}\hat{B}}(t). \qquad (13.3.2)$$

(iii) *Reciprocity Law: If a static magnetic field* **B** *is present, then*

$$\Phi_{\hat{B}\hat{A}}(t, \mathbf{B}) \;=\; \varepsilon_{\hat{A}} \varepsilon_{\hat{B}} \Phi_{\hat{B}\hat{A}}(-t, -\mathbf{B}) \;=\; \varepsilon_{\hat{A}} \varepsilon_{\hat{B}} \Phi_{\hat{A}\hat{B}}(t, -\mathbf{B}). \qquad (13.3.3)$$

Similarly, the Admittance also have the symmetry relations

Theorem 13.3.2

$$\chi_{\hat{B}\hat{A}}(\omega) \overset{\text{def}}{=} \int_0^\infty \Phi_{\hat{B}\hat{A}} e^{-i\omega t} \, dt, \qquad (13.3.4)$$

$$\Re\chi_{\hat{B}\hat{A}}(\omega) = \Re\chi_{\hat{B}\hat{A}}(-\omega), \qquad (13.3.5)$$

$$\Im\chi_{\hat{B}\hat{A}}(\omega) = -\Im\chi_{\hat{B}\hat{A}}(-\omega), \qquad (13.3.6)$$

$$\chi_{\hat{B}\hat{A}}(\omega, -\mathbf{H}) = \varepsilon_{\hat{A}}\varepsilon_{\hat{B}}\chi_{\hat{A}\hat{B}}(\omega, \mathbf{H}). \qquad (13.3.7)$$

Since proofs of these theorems are rather straightforward we refrain from stating them.

13.4   Fluctuation-Dissipation Theorem

We have seen in § 13.2 how the admittance is connected with the equilibrium average of the fluctuation of corresponding currents. It is *not* a special feature of Kubo formalism. Whenever there is a randomization for whatsoever reason, the response of a system to an external perturbation is connected with the thermal fluctuation in the absence of perturbation. In electrical engineering Nyquist's theorem connects frequency-dependent electrical conductivity to the Fourier transform of the thermal average of current-current correlation. The other classic case of such studies is that of the **motion of heavy Brownian particles** in the medium of a dynamical system consisting of large number of degrees of freedom, like the molecules of a liquid.

One-dimentioanal motion of a Brownian particle of mass m in a *dissipative* viscous medium is mathematically described by the **Langevin Equation**

$$\frac{du(t)}{dt} = -\int_{-\infty}^t \gamma(t - t')u(t')dt' + \frac{K(t)}{m} + \frac{R(t)}{m}, \qquad (13.4.1)$$

where

$$\gamma(\tau) = \text{non} - \text{local frictional coefficient}, \qquad (13.4.2)$$

$$K(t) = \text{External Force},$$

$$R(t) = \text{Random Force exerted by the molecules of the medium}.$$

Physical conditions like randomness of $R(t)$, reality of γ and more general causality imposes the following constraints

$$\langle R(t) \rangle = 0 : \qquad \text{randomness of } R, \qquad (13.4.3)$$

$$\gamma(\tau)^* = \gamma(\tau) : \qquad \text{reality of } \gamma, \qquad (13.4.4)$$

$$\gamma(\tau) = 0, \quad \text{for } \tau \leq 0 : \quad \text{causality.} \qquad (13.4.5)$$

We now use the method of *Harmonic Analysis*

$$z(t) = \int_{-\infty}^{+\infty} z(\omega) \exp(+i\omega t) d\omega, \qquad (13.4.6)$$

$$z(\omega) = \frac{1}{2\pi} \int_{-\infty}^{+\infty} z(t) \exp(-i\omega t) dt \qquad (13.4.7)$$

to obtain in the absence of the random force (i.e. $R(t) = 0$) a formal expression of *generalized admittance/mobility* $\mu(\omega)$

$$u(\omega) = \mu(\omega) K(\omega), \qquad (13.4.8)$$

$$\mu(\omega) = \frac{1}{m} \frac{1}{i\omega + \gamma[\omega]}. \qquad (13.4.9)$$

The Laplace-Fourier transform

$$\gamma[\omega] = \int_0^\infty \gamma(\tau) \exp(-i\omega\tau) d\tau \qquad (13.4.10)$$

with

$$\gamma[-\omega^*] = \gamma[\omega]^* \qquad (13.4.11)$$

is analytic in the lower half-plane of the complex ω-plane. The sine/cosine Fourier transforms are given by

$$\Re \gamma[\omega] = \int_0^\infty \gamma(\tau) \cos(\omega\tau) d\tau, \qquad (13.4.12)$$

$$\Im \gamma[\omega] = -\int_0^\infty \gamma(\tau) \sin(\omega\tau). \qquad (13.4.13)$$

By a similar analysis of the force-free case, *i.e.* when $K(t) = 0$, we get

$$u(\omega) = \frac{1}{m} \frac{R(\omega)}{i\omega + \gamma[\omega]}. \qquad (13.4.14)$$

For further progress in this analysis we have the most useful technique of power-spectrum analysis. The **power spectrum**

$$I_z(\omega) = |z(\omega)|^2 = I_z(\omega)^* \geq 0 \qquad (13.4.15)$$

in unit frequency band around frequency ω of a time-signal $z(t)$ is connected
with the **auto-correlation function**

$$\phi_z(t) = \langle z(t_0)^* z(t_0 + t) \rangle \tag{13.4.16}$$

$$= \frac{1}{2\pi} \int_{-\infty}^{+\infty} z(t_0)^* z(t_0 + t) dt_0 \tag{13.4.17}$$

with

$$\phi_z(-t) = \phi_z(t)^* \tag{13.4.18}$$

through Wiener-Khinchin Theorem

$$\phi_z(t) = \int_{-\infty}^{+\infty} I_z(\omega) \exp(i\omega t) d\omega, \tag{13.4.19}$$

$$I_z(\omega) = \frac{1}{2\pi} \int_{-\infty}^{+\infty} \phi_z(t) \exp(-i\omega t) dt. \tag{13.4.20}$$

Applying these results to the velocity $u(t)$ and the random force $R(t)$ we
arrive at

$$I_u(\omega) = \frac{1}{m^2} \frac{I_R(\omega)}{|i\omega + \gamma[\omega]|^2}, \tag{13.4.21}$$

$$\langle u(t_0)u(t_0 + t) \rangle = \frac{1}{m^2} \int_{-\infty}^{+\infty} \frac{I_R(\omega)}{|i\omega + \gamma[\omega]|^2} e^{+i\omega t} d\omega. \tag{13.4.22}$$

Using the contour C shown in Figure 13.4.1 we can evaluate the integral
on the right hand side of Equation 13.4.22 and arrive at

$$\langle u(t_0)u(t_0 + t) \rangle = \frac{1}{m^2} \int_{-\infty}^{+\infty} \frac{1}{i\omega + \gamma[\omega]} \frac{I_R(\omega)}{2\Re\gamma[\omega]} e^{+i\omega t} d\omega \tag{13.4.23}$$

$$= \frac{\pi}{m^2} \frac{I_R(\omega_0)}{\Re\gamma[\omega_0]} e^{-t\gamma[\omega_0]}, \tag{13.4.24}$$

where

$$\omega_0 = -\gamma[\omega_0]. \tag{13.4.25}$$

We now make an assumption which is valid for a large class physical processes, and which is definitely true for Maxwell's distribution. This is the
well-known case of **thermal noise**

$$\langle u(t_0)^2 \rangle = \frac{k_B T}{m}, \tag{13.4.26}$$

$$m\Re\gamma[\omega] = \frac{\pi}{k_B T} I_R(\omega). \tag{13.4.27}$$

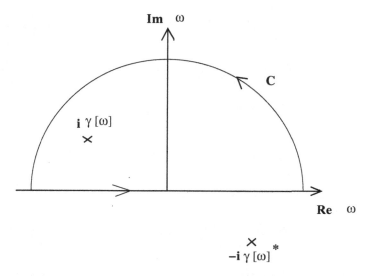

Fig. 13.4.1   Integration contour C and the position of the simple poles of the integrand in Equation 13.4.22 in the complex ω plane.

Using the definition of power spectrum and *stationarity* of the autocorrelation function we arrive at

$$m\Re\gamma[\omega] = \frac{\pi}{k_B T}\frac{1}{2\pi}\int_{-\infty}^{+\infty}\langle R(t_0)R(t_o+t)\rangle e^{-i\omega t}dt$$

$$= \frac{1}{k_B T}\int_0^{+\infty}\langle R(t_0)R(t_0+t)\rangle \cos(\omega t)dt. \qquad (13.4.28)$$

We now apply the well-known result of complex analysis connecting the real and the imaginary parts of an analytic function to obtain the imaginary part of the analytic function $\gamma[\omega]$ and get

$$m\Im\gamma[\omega] = -\frac{1}{k_B T}\int_0^{+\infty}\langle R(t_0)R(t_0+t)\rangle \sin(\omega t)dt \qquad (13.4.29)$$

and finally arrive at

$$m\gamma[\omega] = \frac{1}{k_B T}\int_0^{+\infty}\langle R(t_0)R(t_0+t)\rangle e^{-i\omega t}dt. \qquad (13.4.30)$$

Equation 13.4.30 is the **Fluctuation-Dissipation Theorem of the second kind** connecting the dissipative frictional coefficient with the Fourier transform of the autocorrelation function of the random force.

Referring to Equation 13.4.23 we see that in the case of thermal noise

$$\frac{2\pi}{k_B T}\langle u(t_0)u(t_0 + t)\rangle = \int_{-\infty}^{+\infty} \frac{1}{m}\frac{e^{+i\omega t}}{i\omega + \gamma[\omega]}d\omega. \qquad (13.4.31)$$

Taking inverse Fourier transform we obtain the **Fluctuation-Dissipation Theorem of the first kind**

$$\mu(\omega) = \frac{1}{m}\frac{1}{i\omega + \gamma[\omega]} = \frac{1}{k_B T}\int_{-\infty}^{+\infty}\langle u(t_0)u(t_0 + t)\rangle e^{-i\omega t}dt \qquad (13.4.32)$$

connecting the *generalized admittance/mobility* and the Fourier transform of the *current-current correlation*.

We here add a note on noise. Though **thermal noise** is found in many physical processes, it is not the only type of noise that we encounter. In a wide class of instruments encompassing transistors and vacuum-tubes we come across what is known as **white noise** with the power spectrum $I_R(\omega) =$ constant. The adjective *white* is used from analogy of *white light* in which all frequencies have equal amplitude. A third type of noise known as **pink or 1/f noise** is not only present in electronic instruments but in strange areas like east-west component of ocean current velocity, loudness/pitch fluctuation versus frequency curves of classical radio stations, rock stations and even drum music of african tribes. Its name comes from the power spectrum $I_R(\omega = 2\pi f) \propto 1/f$. Sometimes pink noise has the power spectrum $I_R(\omega) \propto 1/f^\alpha$ with $0.5 \le \alpha \le 1.5$.

13.5   Problems

Problem 13.1.   Prove Equations 13.2.10 and 13.2.12.

Problem 13.2.   Prove Equations 13.2.31 and 13.2.32.

Problem 13.3.   Prove Equation 13.2.45.

Problem 13.4.   Prove Equations 13.2.47 and 13.2.48.

Problem 13.5.   Prove Equations 13.3.1, 13.3.2 and 13.3.3.

Problem 13.6.   Prove Equations 13.3.4, 13.3.5, 13.3.6 and 13.3.7.

Problem 13.7.   Prove Equations 13.4.21 and 13.4.22.

Problem 13.8.   Prove Equations 13.4.30.

Chapter 14

Mathematical Appendix

14.1 Beta and Gamma Functions

In calculating different physical quantities, the gamma and the beta functions are used so many times that we have collected here the important properties of these functions.

Defining Relations:

$$\Gamma(z) = \int_0^\infty e^{-t} t^{z-1} dt, \tag{14.1.1}$$

$$B(m, n) = \int_0^1 x^{m-1}(1-x)^{n-1} dx \tag{14.1.2}$$

$$= 2 \int_0^{\pi/2} \sin^{2m-1}(\theta) \cos^{2n-1}(\theta) d\theta. \tag{14.1.3}$$

Some Important Relations:

$$\Gamma(z+1) = z\Gamma(z), \tag{14.1.4}$$

$$\Gamma\left(\frac{1}{2}\right) = \sqrt{\pi}, \tag{14.1.5}$$

$$\Gamma(1) = 1, \tag{14.1.6}$$

$$\Gamma(n+1) = n!, \text{ for positive integral } n, \tag{14.1.7}$$

$$B(m, n) = B(n, m), \tag{14.1.8}$$

$$B(m, n) = \frac{\Gamma(m)\Gamma(n)}{\Gamma(m+n)}. \tag{14.1.9}$$

In Figure 14.1.1 $\Gamma(t)$ as a function of the real variable t has been shown. It has got singularities at t = 0, -1, -2, $\cdots$. For the positive values of t, it has a minimum between +1 and +2.

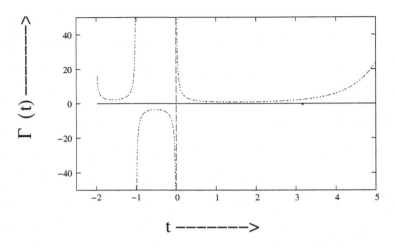

Fig. 14.1.1   Form of $\Gamma(t)$ as a function of the real variable t.

14.2   Dirac Delta Function (Distribution)

This is one other function that a physicist cannot do without and here is the summary of the most important properties.

Defining Relations for a single variable:

$$\delta(x) = \delta(-x), \tag{14.2.1}$$

$$x \times \delta(x) = 0, \tag{14.2.2}$$

$$\int_{a-\eta}^{a+\eta} f(x)\delta(x-a)\,dx = f(a) \ \text{ for all real } a \ \& \ \text{ real positive } \eta. \tag{14.2.3}$$

Different Representations:

$$\delta(x) = \frac{1}{2\pi}\int_{-\infty}^{+\infty} \exp\,(\pm ixt)\ dt. \tag{14.2.4}$$

For any function f(x) satisfying

$$\int_{-\infty}^{+\infty} f(x)dx = 1, \tag{14.2.5}$$

$$\delta(x) = \lim_{a\to\infty} af(ax). \tag{14.2.6}$$

Some Important Properties:

$$\int_{-\infty}^{+\infty} f(x)\delta'(x-a)dx = -f'(a) \quad \text{for all real } a, \qquad (14.2.7)$$

$$\delta((f(x))) = \sum_{i=1}^{i=n} \frac{\delta(x-a_i)}{|f'(a_i)|}, \qquad (14.2.8)$$

where a_i, $i = 1, 2, \cdots, n$ are the n simple roots of $f(x) = 0$.
Connection with **Heaviside Θ-function**:

$$\Theta(x-a) = \begin{cases} 0, & \text{if x } < \text{ a} \\ 1, & \text{if x } > \text{ a} \end{cases} \qquad (14.2.9)$$

$$\delta(x-a) = \Theta'(x-a). \qquad (14.2.10)$$

We have plotted Heaviside's Θ-function in Figure 14.2.1.

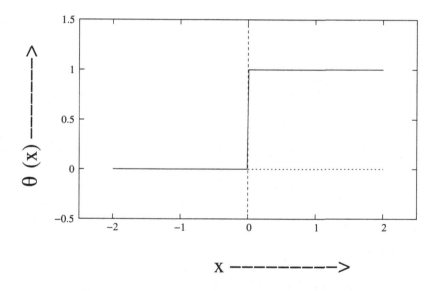

Fig. 14.2.1 Heaviside's Θ-function as a function of x.

Forms for 3-dimensional space:

$$Cartesian\ Coordinates: \quad \delta(\mathbf{r} = (x, y, z)) = \delta(x)\ \delta(y)\ \delta(z),$$

$$(14.2.11)$$

$$Spherical\ Coordinates: \quad \delta(\mathbf{r} = (r, \theta, \phi)) = \frac{\delta(r)}{2\pi\ r^2}, \quad (14.2.12)$$

$$Cylindrical\ Coordinates: \quad \delta(\mathbf{r} = (r, \theta, z)) = \frac{\delta(r)\delta(z)}{\pi\ r}. \quad (14.2.13)$$

14.3   Functional Derivative

A **functional** $F[\phi]$ of a function $\phi(x)$ is defined as

$$F[\phi] = \int F\left(\phi(x), \phi_1(x), \phi_2(x), \cdots\right) dx, \qquad (14.3.1)$$

where

$$\phi_k(x) = \frac{\partial^k}{\partial x^k} \phi(x). \qquad (14.3.2)$$

We now define **functional derivative** by the relation

$$\int \frac{\delta F[\phi]}{\delta \phi} f(x) dx = \lim_{\epsilon \to 0} \frac{F[\phi + \epsilon f] - F[\phi]}{\epsilon}. \qquad (14.3.3)$$

We continue the calculation

$$\int \frac{\delta F[\phi]}{\delta \phi} f(x) dx = \lim_{\epsilon \to 0} \frac{F[\phi + \epsilon f] - F[\phi]}{\epsilon}$$

$$= \lim_{\epsilon \to 0} \frac{1}{\epsilon} \int \left[F\left(\phi(x) + \epsilon f(x), \phi_1(x) + \epsilon f_1(x), \cdots\right) \right.$$

$$\left. - F\left(\phi(x), \phi_1(x), \phi_2(x), \cdots\right) \right] dx$$

$$= \int \sum_{k=0}^{\infty} \frac{\partial F}{\partial \phi_k} f_k(x) dx \;\; = \;\; \int \left[\sum_{k=0}^{\infty} (-1)^k \left(\frac{d}{dx}\right)^k \frac{\partial F}{\partial \phi_k} \right] f(x) dx.$$

The last step follows if we use
$\int u(x) \left(\frac{d}{dx}\right)^k v(x) dx = (-1)^k \int \left(\left(\frac{d}{dx}\right)^k u(x) \right) v(x) dx$
and suitable boundary conditions. We now obtain the expression for calculating the **functional derivative**

$$\frac{\delta F[\phi]}{\delta \phi} = \sum_{k=0}^{\infty} (-1)^k \left(\frac{d}{dx}\right)^k \frac{\partial F}{\partial \phi_k}. \qquad (14.3.4)$$

14.4 Mathematical Identities

Theorem 14.4.1. $e^{\lambda S} A e^{-\lambda S} = \sum_{m=0}^{\infty} \frac{\lambda^m}{m!} [S, [S, \cdots, [S, A] \cdots]]_{m \text{ terms}}.$

Proof: We note that

$$
e^{\lambda S} A e^{-\lambda S} = \sum_{m,n=0}^{\infty} (-1)^n \frac{\lambda^{m+n}}{m!\, n!} S^m A S^n
$$

$$
= \sum_{m=0}^{\infty} \frac{\lambda^m}{m!} \sum_{k=0}^{m} (-1)^k \binom{m}{k} S^{m-k} A S^k.
$$

We shall now show by the Method of Transfinite Induction that

$$
\sum_{k=0}^{m} (-1)^k \binom{m}{k} S^{m-k} A S^k = [S, [S, \cdots, [S, A] \cdots]]_{m \text{ terms}}. \qquad (14.4.1)
$$

Equation 14.4.1 is of course true for m = 1. We assume this equation to be true for m = M, i.e.

$$
\sum_{k=0}^{M} (-1)^k \binom{M}{k} S^{M-k} A S^k = [S, [S, \cdots, [S, A] \cdots]]_{M \text{ terms}}. \qquad (14.4.2)
$$

We can now calculate

$$
\sum_{k=0}^{M+1} (-1)^k \binom{M+1}{k} S^{M+1-k} A S^k = \sum_{k=0}^{M+1} (-1)^k \left(\binom{M}{k} + \binom{M}{k-1} \right)
$$

$$
\times S^{M+1-k} A S^k
$$

$$
= S \left(\sum_{k=0}^{M} (-1)^k \binom{M}{k} S^{M-k} A S^k \right)
$$

$$
- \left(\sum_{k=0}^{M} (-1)^k \binom{M}{k} S^{M-k} A S^k \right) S
$$

$$
= S [S, [S, \cdots, [S, A] \cdots]]_{M \text{ terms}}
$$

$$
- [S, [S, \cdots, [S, A] \cdots]]_{M \text{ terms}} S
$$

$$
= [S, [S, \cdots, [S, A] \cdots]]_{M+1 \text{ terms}}.
$$

Equation 14.4.1 is thus universally true and the Theorem 14.4.1 is thus proved.

14.5   Multiple Summation

Here we prove a result of multiple summation

$$\sum_{N=0}^{\infty} \sum_{n_1+n_2+\cdots=N} f(n_1, n_2, \cdots) = \sum_{n_1=0}^{\infty} \sum_{n_2=0}^{\infty} \cdots f(n_1, n_2, \cdots). \qquad (14.5.1)$$

For a function with 2 indices, $f(n_1, n_2)$ the summation on the left hand side of Equation 14.5.1 is successively evaluated for the grid points lying on different slanted lines shown in Figure 14.5.1. The summation is thus

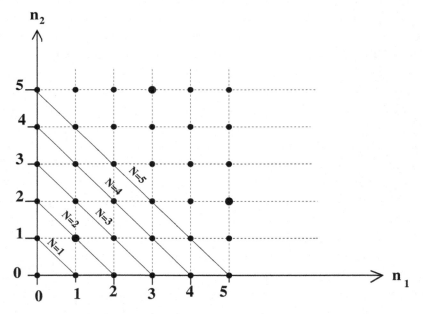

Fig. 14.5.1   Grid points for a 2-indexed function $f(n_1, n_2)$ and the slanted lines along which $n_1 + n_2 = N$ is constant.

evaluated for all the points shown in the figure. This is exactly the summation on the right hand side of the Equaion. The general result follows from application of the Method of Transfinite Induction.

14.6 Pauli Matrices

Like in many other branches of physics the 2-level spin-$\frac{1}{2}$ system is a very convenient example in Statistical Physics. In treating this system the Pauli matrices are unavoidable.

$$\sigma_x = \begin{pmatrix} 0 & 1 \\ 1 & 0 \end{pmatrix}, \tag{14.6.1}$$

$$\sigma_y = \begin{pmatrix} 0 & -i \\ i & 0 \end{pmatrix}, \tag{14.6.2}$$

$$\sigma_z = \begin{pmatrix} 1 & 0 \\ 0 & -1 \end{pmatrix}. \tag{14.6.3}$$

Properties of the Pauli Matrices:

$$\sigma_\mu \sigma_\nu = \delta_{\mu\nu} + i \sum_\lambda e_{\mu\nu\lambda} \sigma_\lambda, \tag{14.6.4}$$

$$(\vec{\sigma} \cdot \mathbf{a})(\vec{\sigma} \cdot \mathbf{b}) = (\mathbf{a} \cdot \mathbf{b}) + i\vec{\sigma} \cdot (\mathbf{a} \times \mathbf{b}). \tag{14.6.5}$$

Here $e_{\mu\nu\lambda}$ denotes the 3-dimensional **Permutation Symbol** having the very important property

$$\sum_\lambda e_{\alpha\beta\lambda} e_{\mu\nu\lambda} = \delta_{\alpha\mu}\delta_{\beta\nu} - \delta_{\alpha\nu}\delta_{\beta\mu}. \tag{14.6.6}$$

14.7 Probability Theory

Statistical Physics being essentially probabilistic in nature it is worth knowing the basic results of **Mathematical Probability**.

14.7.1 *Elementary Results of Probability Theory*

Some of the results of Probability Theory are given here.

Theorem 14.7.1 (of Total Probability).
$Pr\{A \bigcup B\} = Pr\{A\} + Pr\{B\} - Pr\{A \bigcap B\}$.

Theorem 14.7.2 (of Compound Probability).
$Pr\{A \bigcap B\} = Pr\{A\}Pr\{B|A\}$.

Here for any two events B and A, $Pr\{B|A\}$ means the **conditional probability** of occurrence of the event B, given that the event A has occurred.

In Mathematical Statistics, two events B and A are called *statistically independent* if $Pr\{B|A\} = Pr\{B\}$. Thus we see that the concept of *statistical independence* is the same in both the disciplines.

Theorem 14.7.3 (Bayes'). $Pr\{B_i|A\} = \frac{Pr\{B_i\}Pr\{A|B_i\}}{\sum_k Pr\{B_k\}Pr\{A|B_k\}}.$

14.7.2　*Statistical Distributions*

A number of these statistical distributions frequently appear in Physics and we now describe them.

(14.7.2.1)　*Binomial/Bernoulli Distribution*

This is the most ubiquitious of all the distributions:

$$W_B(n, N; p) = \frac{\Gamma(N+1)}{\Gamma(n+1)\,\Gamma(N-n+1)} p^n (1-p)^{N-n}. \qquad (14.7.1)$$

Physicists call it **Binomial Distribution**, while to mathematicians it is **Bernoulli Distribution**. Figure 14.7.1 depicts the distribution for differ-

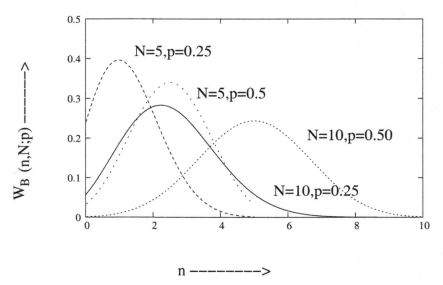

Fig. 14.7.1　Plot of $W_B(n, N; 0 \le p \le 1) = \frac{\Gamma(N+1)}{\Gamma(n+1)\Gamma(N-n+1)} p^n (1-p)^{N-n}$ versus n for different values of N and p.

ent values of the parameters. For discrete values of n the mean m and the variance σ^2 are

$$m = \sum_{n=0}^{N} nW_B(n, N; p) = \sum_{n=0}^{N} n\frac{N!}{n!(N-n)!}p^n(1-p)^{N-n}$$
$$= Np, \tag{14.7.2}$$

$$\sigma^2 = \sum_{n=0}^{N} n^2 W_B(n, N; p) = \sum_{n=0}^{N} n^2\frac{N!}{n!(N-n)!}p^n(1-p)^{N-n}$$
$$= Np(1-p). \tag{14.7.3}$$

Three examples from different physical systems are:

(i) Gas in a Container,
(ii) Spin 1/2 particles,
(iii) Random Walk.

(14.7.2.2) *Poisson Distribution*

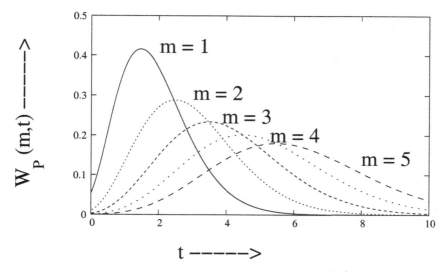

Fig. 14.7.2   Form of the Poisson Distribution $W_P(m, t) = \frac{e^{-m}m^t}{\Gamma(t+1)}$ as a function of t for different values of m.

Error analysis in nuclear radioactive decays satisfies Poisson distribution:

$$W_P(m, t) = \frac{e^{-m}m^t}{\Gamma(t+1)}. \tag{14.7.4}$$

In Figure 14.7.2 we have plotted the Poisson Distribution for different values of m. Poisson distribution can be obtained as a limiting case of Binomial distribution:

$$W_P(m,t) = \lim_{\substack{p\to 0, N\to\infty \\ Np=m}} W_B(t,N;p). \tag{14.7.5}$$

Thus the *mean* equals m and the *variance* also equals m. This explains why in nuclear decay experiments the error bar is taken as the square root of the expermental value.

(14.7.2.3) *Gaussian/Normal Distribution*

What is called the **Gaussian Distribution**, with mean $\bar{x}$ and standard deviation σ, by the physicists is the **Normal distribution** to the mathematicians:

$$W_G(x,\bar{x},\sigma) = \frac{1}{\sqrt{2\pi}\sigma} e^{-\frac{(x-\bar{x})^2}{2\sigma^2}}. \tag{14.7.6}$$

This distribution is graphically presented in Figure 14.7.3 The importance

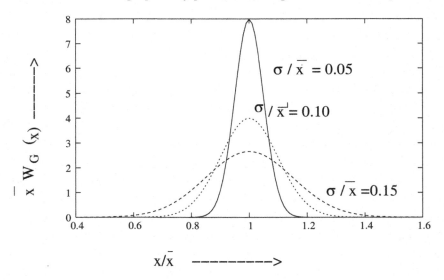

Fig. 14.7.3   Plot of $\bar{x}W_G(x,\bar{x},\sigma)$ as a function of $\frac{x}{\bar{x}}$ for different values of $\frac{\sigma}{\bar{x}}$.

of this distribution (and hence its epithet *Normal*) for macroscopic bodies with number of particles $\sim 10^{23}$ becomes clear from Theorem 14.7.4. Also spectral lines from solids have Gaussian profiles.

(14.7.2.4) *Lorentzian/Cauchy Distribution*

Spectral lines from gases and liquids have Lorentzian (Cauchy in the language of mathematicians) profiles, shown in Figure 14.7.4. The Lorentzian

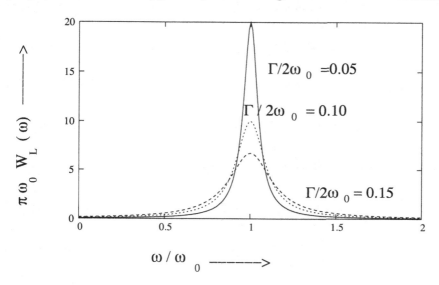

Fig. 14.7.4   Plot of $\pi\omega_0 W_L\left(\omega, \omega_0, \Gamma\right)$ as functions of $\frac{\omega}{\omega_0}$ for different values of $\frac{\Gamma}{2\omega_0}$.

dustribution

$$W_L\left(\omega, \omega_0, \Gamma\right) = \frac{\frac{\Gamma}{2\pi}}{\left(\omega - \omega_0\right)^2 + \left(\frac{\Gamma}{2}\right)^2} \tag{14.7.7}$$

has the intersting feature that all its moments diverge.

The contrasting features of the Lorentzian and the Gaussian distribuions have to be noted. The Lorentzian distribution has a very sharp peak at the mean but a very long tail; while the Gaussian distribution has a broad peak but a short tail.

14.7.3   *Central Limit Theorem*

Theorem 14.7.4   *If the probability distribution W(n,N) of the stochastic variable n for a N-particle system satisfies the following conditions:*

(i) W(n,N) has a sharp maximum at n_0,
(ii)

$$\frac{d^{k+1}}{dn_0^{k+1}} \ln W(n, N) \propto \frac{1}{N} \frac{d^k}{dn_0^k} \ln W(n, N),$$

then in the limit $N \to \infty$, W(n,N) tends to the Gaussian (Normal) distribution

$$W_G(n) = \frac{1}{\sqrt{2\pi}\sigma} e^{-\frac{1}{2\sigma^2}(n-\bar{n})^2},$$

where $\bar{n}$ is the mean of n and σ is the standard deviation of n.

Proof: Since $\ln W(n)$ is a monotonous function of W(n), the conditions of the Theorem allow us to write

$$\ln W(n, N) = \ln W(n_0, N) - \frac{1}{2!}|B_2|\eta^2 + \frac{1}{3!}B_3\eta^3 + \cdots , \qquad \text{where}$$

$$\eta = n - n_0, \qquad \text{and}$$

$$B_k = \frac{d^k}{dn_0^k} \ln W(n, N).$$

We thus get

$$\lim_{N\to\infty} W(n, N) = \lim_{N\to\infty} W(n_0, N) e^{-\frac{1}{2}|B_2|\eta^2}.$$

The Theorem is thus proved.

Definitions of mean, variance, and normalization of W(n,N) allow us to obtain

$$\bar{n} \equiv \sum_{n=0}^{\infty} n \lim_{N\to\infty} W(n, N) = n_0,$$

$$\sigma^2 \equiv \sum_{n=0}^{\infty} (n - \bar{n})^2 \lim_{N\to\infty} W(n, N) = \frac{1}{|B_2|}.$$

In Figure 14.7.5 we show how a particular distribution with parameter N tends to the Gaussian (Normal) distribution as N increases. It is to be remembered that the Gaussian distribution is a *platykurtic* curve.

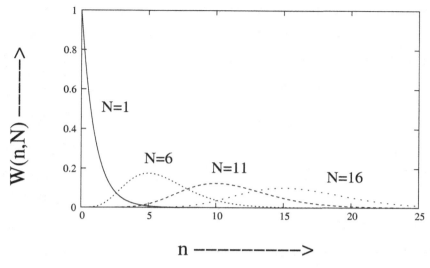

Fig. 14.7.5 Form of $W(n, N) = e^{-n} n^{N-1}/\Gamma(N)$ as a function of n for different values of N.

14.8 Quantum Mechanics, A Retrospect

In this section we summarize the notations and methods of quantum mechanical states and operators that we shall use in the text.

$$\text{Basis Set}: |n\rangle, \tag{14.8.1}$$

$$\text{Orthonormality}: \langle m|n\rangle = \delta_{m,n}, \tag{14.8.2}$$

$$\text{Completeness}: \sum_n |n\rangle\langle n| = \hat{I}, \text{ the identity operator,} \tag{14.8.3}$$

$$\text{Quantum Mechanical State}: |\psi(t)\rangle = \sum_n C_n(t)|n\rangle, \tag{14.8.4}$$

$$\text{Quantum Mechanical Operator}: \hat{A} = \sum_{m,n} |m\rangle A_{m,n} \langle n|, \tag{14.8.5}$$

$$\text{Expectation Value of Operator}: \langle \hat{A} \rangle = \langle \psi(t)|\hat{A}|\psi(t)\rangle$$
$$= \sum_{m,n} C_m(t)^* A_{m,n} C_n(t), \tag{14.8.6}$$

$$\text{where } A_{m,n} = \langle m|\hat{A}|n\rangle. \tag{14.8.7}$$

In Coordinate Representation:

$$\text{Basis Function}: \quad \phi_n(\mathbf{r}) = \langle \mathbf{r}|n\rangle, \tag{14.8.8}$$

$$\text{State Function}: \quad \psi(\mathbf{r},t) = \sum_n C_n(t)\phi_n(\mathbf{r}), \tag{14.8.9}$$

$$\text{Operator}: \quad \hat{A}(\mathbf{r'},\mathbf{r}) = \sum_{m,n} \phi_m(\mathbf{r'})A_{m,n}\phi_n(\mathbf{r})^*. \tag{14.8.10}$$

The **Interaction (Dirac)** and the **Heisenberg** pictures of operators are given by:

$$\hat{A}_D(t) = e^{+i\frac{\hat{\mathcal{H}}_0 t}{\hbar}}\hat{A}e^{-i\frac{\hat{\mathcal{H}}_0 t}{\hbar}}, \tag{14.8.11}$$

$$\hat{A}_H(t) = e^{+i\frac{\hat{\mathcal{H}} t}{\hbar}}\hat{A}e^{-i\frac{\hat{\mathcal{H}} t}{\hbar}}. \tag{14.8.12}$$

Here $\hat{\mathcal{H}}$ and $\hat{\mathcal{H}}_0$ are the total and the unperturbed *hamiltonians* of the system.

Equation of Motion in Heisenberg picture is

$$i\hbar\dot{\hat{A}}_H = i\hbar\frac{\partial \hat{A}_H}{\partial t} + \left[\hat{A}_H, \hat{\mathcal{H}}\right]. \tag{14.8.13}$$

Equation of Motion in Dirac picture is

$$i\hbar\dot{\hat{A}}_D = i\hbar\frac{\partial \hat{A}_D}{\partial t} + \left[\hat{A}_D, \hat{\mathcal{H}}_0\right]. \tag{14.8.14}$$

14.9　Riemann, Bernoulli and Fourier

14.9.1　*Riemann's ζ-Function*

In almost all calculations in Statistical Physics use of the Riemann's ζ-function makes algebraic manipulation very easy. Defining Relation:

$$\zeta(a) = \sum_{n=1}^{\infty} \frac{1}{n^a}, \tag{14.9.1}$$

uniformly convergent if & only if $\text{Re}(a) > 1$.

Some Important Properties:

(i) For $\text{Re}(a) \leq 0$, $\zeta(a)$ has zeroes only at $a = -2, -4, \cdots$.

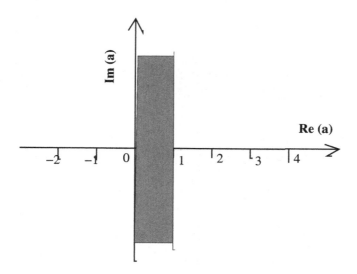

Fig. 14.9.1   Region of complex-a plane with Re(a)$\geq$ 0 where $\zeta(a)$ has a zero.

(ii) **Riemann's Conjecture:** For Re(a)$>$ 0, the zeroes of ζ(a) lie only within the strip $0 \leq$ Re(a) ≤ 1, shown in Figure 14.9.1.

(iii) $2^{1-a}\Gamma(a)\zeta(a)\cos\left(\frac{a\pi}{2}\right) = \pi^a\zeta(1-a)$.

(iv) **Euler's Relation:** $\frac{1}{\zeta(a)} = \prod_{p=\text{primes}}\left(1 - \frac{1}{p^a}\right)$.

Some of the important ζ-functions, needed in Statistical Physics, are:

$$\zeta\left(\frac{3}{2}\right) = 2.612\cdots, \ \zeta\left(\frac{5}{2}\right) = 1.342\cdots, \ \zeta(2) = \frac{\pi^2}{6}.$$

14.9.2   *Bernoulli Numbers and Polynomials*

The expressions in Statistical Physics sometimes use the Bernoulli Numbers instead of the Riemannian ζ-functions.

Generating Function for Bernoulli Polynomial:

$$\frac{te^{tx}}{e^t - 1} = \sum_{n=0}^{\infty} B_n(x)t^n. \qquad (14.9.2)$$

Special Values: $B_0(x) = 1, B_1(x) = x - \frac{1}{2}, B_2(x) = x^2 - x + \frac{1}{6}$.
Generating Function for Bernoulli Numbers:

$$\frac{x}{e^x - 1} = \sum_{n=0}^{\infty} \frac{B_n x^n}{\Gamma(n+1)}. \qquad (14.9.3)$$

Special Values: $B_0 = 1, B_1 = -\frac{1}{2}, B_2 = \frac{1}{6}, B_4 = -\frac{1}{30}$.
Connection with ζ-function:

$$\zeta(2m) = \frac{2^{2m-1}\pi^{2m}B_{2m}}{\Gamma(2m+1)}, \quad m = 0, 1, 2, 3, \cdots. \qquad (14.9.4)$$

14.9.3   *Fourier Series*

Evaluation of Riemannian ζ-functions for positive even integral arguments can easily be done by using Fourier Series.

Theorem 14.9.1. (Fourier's).
A function $f(x)$ can be expanded in a series in the region $[-L \leq x \leq +L]$:

$$f(x) = \frac{a_0}{2} + \sum_{n=1}^{\infty} \left(a_n \cos\frac{n\pi x}{L} + b_n \sin\frac{n\pi x}{L} \right) \qquad (14.9.5)$$

with

$$a_n = \frac{1}{L} \int_{-L}^{+L} f(x)\cos\frac{n\pi x}{L}dx, \quad n = 0, 1, 2, \cdots \qquad (14.9.6)$$

and

$$b_n = \frac{1}{L} \int_{-L}^{+L} f(x)\sin\frac{n\pi x}{L}dx, \quad n = 1, 2, 3, \cdots, \qquad (14.9.7)$$

if in any finite domain f(x) satisfies the following **Dirichlet's Conditions:**

  (i) f(x) is continous except for a finite number of discontinuities of finite magnitude,
  (ii) f(x) has a finite number of extrema.

The most famous example of the function that **cannot** be expanded in a Fourier Series is $f(x) = \sin \frac{1}{x}$. This function has been plotted in Figure 14.9.2.

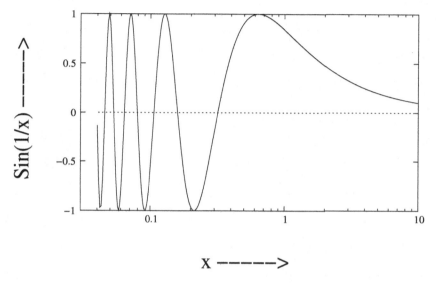

Fig. 14.9.2   Plot of $\sin \frac{1}{x}$ as a function of x.

The method of Fourier Series can be utilized to evaluate ζ-functions with positive even integral arguments, $\zeta(2m)$, $m = 1, 2, \cdots$. As an example, for m=1, we obtain:

$$f(x) = x^2, \tag{14.9.8}$$

$$a_0 = \frac{2}{L} \int_0^L x^2 dx = \frac{2L^2}{3}, \tag{14.9.9}$$

$$a_n = \frac{2}{L} \int_0^L x^2 \cos\left(\frac{n\pi x}{L}\right) dx = (-1)^n \frac{4L^2}{n^2 \pi^2}, \quad \text{for } n \geq 1. \tag{14.9.10}$$

Thus,

$$x^2 = \frac{L^2}{3} + \frac{4L^2}{\pi^2} \sum_{n=1}^{\infty} (-1)^n \frac{\cos\left(\frac{n\pi x}{L}\right)}{n^2}. \tag{14.9.11}$$

Putting x = L in Equation 14.9.11, we get

$$\zeta(2) = \frac{\pi^2}{6}. \tag{14.9.12}$$

14.9.4   *Integrals of Quantum Statistics*

While discussing Quantum Statistical systems we come across integrals of the form $\int_0^\infty \frac{\epsilon^s}{e^{(\epsilon-\mu)/(k_BT)} \mp 1} d\epsilon$ which can be tranformed into integrals of the form
$\int_0^\infty \frac{z^s}{e^{z-\mu/(k_BT)} \mp 1} dz$ and for this integral we obtain (for $\mu \ll k_BT$):

$$
\begin{aligned}
I_{\mp}(s, \mu) &= \int_0^\infty \frac{z^s}{e^{z-\mu/(k_BT)} \mp 1} dz \\
&= \int_0^\infty e^{-z+\mu/(k_BT)} z^s \left[1 \mp e^{-z+\mu/(k_BT)}\right]^{-1} dz \\
&= e^{\mu/(k_BT)} \int_0^\infty e^{-z} z^s \sum_{n=0}^{\infty} (\pm 1)^n e^{-nz+n\mu/(k_BT)} dz \\
&= \sum_{n=0}^{\infty} (\pm 1)^n e^{(n+1)\mu/(k_BT)} \int_0^\infty e^{-(n+1)z} z^s dz \\
&= \sum_{n=0}^{\infty} (\pm 1)^n \frac{e^{(n+1)\mu/(k_BT)}}{(n+1)^{s+1}} \int_0^\infty e^{-t} t^s dt \\
&= \Gamma(s+1) \sum_{n=1}^{\infty} (\pm 1)^{n-1} \frac{e^{n\mu/(k_BT)}}{n^{s+1}}. \tag{14.9.13}
\end{aligned}
$$

For a Bose system below the condensation temperature, $\mu = 0$ and hence

$$
\begin{aligned}
I_-(s, \mu = 0) &= \Gamma(s+1) \sum_{n=1}^{\infty} \frac{1}{n^{s+1}} \\
&= \Gamma(s+1)\zeta(s+1). \tag{14.9.14}
\end{aligned}
$$

For degenerate Fermi systems with $\mu > \epsilon_0$, the lowest value of energy and $k_BT \ll \mu, \mu - \epsilon_0$ the integrals that appear are of the form:

$$I = \int_{\epsilon_0}^{\infty} \frac{f(\epsilon)}{e^{(\epsilon-\mu)/(k_BT)} + 1} d\epsilon$$

$$= \int_{\epsilon_0}^{\mu} \frac{f(\epsilon)}{e^{(\epsilon-\mu)/(k_BT)} + 1} d\epsilon + \int_{\mu}^{\infty} \frac{f(\epsilon)}{e^{(\epsilon-\mu)/(k_BT)} + 1} d\epsilon$$

$$= \int_{\epsilon_0}^{\mu} f(\epsilon) \left[1 - \frac{e^{(\epsilon-\mu)/(k_BT)}}{e^{(\epsilon-\mu)/(k_BT)} + 1} \right] d\epsilon + \int_{\mu}^{\infty} \frac{f(\epsilon)}{e^{(\epsilon-\mu)/(k_BT)} + 1} d\epsilon$$

$$= \int_{\epsilon_0}^{\mu} f(\epsilon) d\epsilon - \int_{\epsilon_0}^{\mu} f(\epsilon) \frac{1}{e^{-(\epsilon-\mu)/(k_BT)} + 1} d\epsilon + \int_{\mu}^{\infty} \frac{f(\epsilon)}{e^{(\epsilon-\mu)/(k_BT)} + 1} d\epsilon$$

$$= \int_{\epsilon_0}^{\mu} f(\epsilon) d\epsilon + k_BT \int_{(\mu-\epsilon_0)/(k_BT)}^{0} \frac{f(\mu - k_BTz)}{e^z + 1} dz$$

$$+ k_BT \int_{0}^{\infty} \frac{f(\mu + k_BTz)}{e^z + 1} dz$$

$$= \int_{\epsilon_0}^{\mu} f(\epsilon) d\epsilon - k_BT \int_{0}^{(\mu-\epsilon_0)/(k_BT)} \frac{f(\mu - k_BTz)}{e^z + 1} dz$$

$$+ k_BT \int_{0}^{\infty} \frac{f(\mu + k_BTz)}{e^z + 1} dz$$

$$\approx \int_{\epsilon_0}^{\mu} f(\epsilon) d\epsilon$$

$$+ k_BT \int_{0}^{\infty} \frac{[f(\mu + k_BTz) - f(\mu - k_BTz)]}{e^z + 1} dz$$

$$= \int_{\epsilon_0}^{\mu} f(\epsilon) d\epsilon + 2 \sum_{n=0}^{\infty} (k_BT)^{2n+2} \frac{f^{(2n+1)}(\mu)}{\Gamma(2n+2)} I_+(2n+1, \mu = 0). \quad (14.9.15)$$

Here we have used Equation 14.9.13 and $f^{(k)}(\mu)$ denotes the k-th derivative of f(x) evaluated at $x = \mu$. To get a complete value of I we note that:

$$I_+(s, \mu = 0) = \Gamma(s+1) \sum_{n=1}^{\infty} \frac{(-1)^{n-1}}{n^{s+1}}$$

$$= \Gamma(s+1) \left[\sum_{n=1}^{\infty} \frac{1}{n^{s+1}} - 2 \sum_{m=1}^{\infty} \frac{1}{(2m)^{s+1}} \right]$$

$$= \Gamma(s+1) \left(1 - \frac{1}{2^s} \right) \sum_{n=1}^{\infty} \frac{1}{n^{s+1}}$$

$$= \Gamma(s+1) \left(1 - \frac{1}{2^s} \right) \zeta(s+1). \quad (14.9.16)$$

Thus the integral finally becomes:

$$I = \int_0^{\mu} f(\epsilon)d\epsilon + 2\sum_{n=0}^{\infty}(k_B T)^{2n+2} f^{(2n+1)}(\mu)\left(1 - \frac{1}{2^{2n+1}}\right)\zeta(2n+2).$$

$$(14.9.17)$$

14.10   Sanskrit Transliteration

We give here the rule suggested by International Alphabet for Sanskrit Transliteration (IAST) for writing Indian words in Roman letters of alphabet.

Table 14.10.1   Rule for writing words of Indian languages in Roman letters of alphabet according to International Alphabet for Sanskrit Transliteration (IAST).

अ = a,   आ = ā,   इ = i,   ई = ī,

उ = u,   ऊ = ū,   ऋ = r̥,   ऌ = l̥,

ए = e,   ऐ = ai,   ओ = o,   औ = au,

क = k,   ख = kh,   ग = g,   घ = gh,   ङ = ṅ,

च = c,   छ = ch,   ज = j,   झ = jh,   ञ = ñ,

ट = ṭ,   ठ = ṭh,   ड = ḍ,   ढ = ḍh,   ण = ṇ,

त = t,   थ = th,   द = d,   ध = dh,   न = n,

प = p,   फ = ph,   ब = b,   भ = bh,   म = m,

य = y,   र = r,   ल = l,   व = v,

श = ś,   ष = ṣ,   स = s,   ह = h,

(˚) = ṁ,   ः = ḥ,   ऽ = (')

14.11 Stirling's Theorem

For discussing statistical physics of macroscopic bodies with the number of particles $\sim 10^{23}$ we need the limiting value of $n!$ and the following theorem is always invoked.

Theorem 14.11.1 (Stirling's).

$$\lim_{n \to \infty} n! = \frac{1}{\sqrt{2\pi}} e^{-n} n^{n+\frac{1}{2}} \left[1 + \frac{1}{12n} + \cdots \right]. \qquad (14.11.1)$$

Proof:

$$\Gamma(n+1) = n! = \int_0^\infty e^{-x} x^n \, dx = \int_0^\infty F(x) dx.$$

The integrand F(x), shown schematicaly in Figure 14.11.1, has a sharp maximum at x = n and so has $\ln F(x)$. Using the new variable $\xi = x - n$,

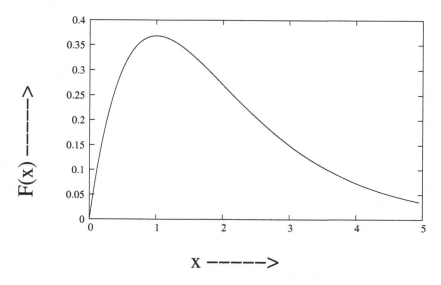

Fig. 14.11.1   Form of F(x) as a function of the positive real variable x, for $n = 1$.

we can write

$$\ln F(x) = n \ln(n + \xi) - (n + \xi)$$

$$= n \ln n - n - \frac{\xi^2}{2n} + \frac{\xi^3}{3n^2} + \frac{\xi^4}{4n^3} + \cdots,$$

$$F(x) = e^{-n} \, n^n \, e^{-\frac{\xi^2}{2n}} \exp\left[\frac{\xi^3}{3n^2} - \frac{\xi^4}{4n^3} + \cdots\right],$$

$$n! = e^{-n} n^n \int_{-n}^{\infty} e^{\frac{-\xi^2}{2n}} \exp\left[\frac{\xi^3}{3n^2} - \frac{\xi^4}{4n^3} + \cdots\right] d\xi,$$

$$\lim_{n \to \infty} n! = e^{-n} n^n \int_{-\infty}^{\infty} e^{\frac{-\xi^2}{2n}} \exp\left[\frac{\xi^3}{3n^2} - \frac{\xi^4}{4n^3} + \cdots\right] d\xi$$

$$= e^{-n} n^n \int_{-\infty}^{\infty} e^{\frac{-\xi^2}{2n}} \left[1 + \left(\frac{\xi^3}{3n^2} - \frac{\xi^4}{4n^3} + \cdots\right)\right.$$

$$+ \frac{1}{2} \left(\frac{\xi^3}{3n^2} - \frac{\xi^4}{4n^3} + \cdots\right)^2$$

$$\left. + \frac{1}{6} \left(\frac{\xi^3}{3n^2} - \frac{\xi^4}{4n^3} + \cdots\right)^3 + \cdots\right] d\xi$$

$$= 2e^{-n} n^n \int_{0}^{\infty} e^{\frac{-\xi^2}{2n}} \left[1 - \frac{\xi^4}{4n^3} + \frac{\xi^6}{18n^4} + \cdots\right] d\xi. \qquad (14.11.2)$$

We retain all those terms in the integrand so that in the final result none of the terms $O(\frac{1}{n})$ are missed. After evaluation of the integrals we arrive at the theorem.

To have a feeling how well the aproximation works we have shown in Table 14.11.1 $\log x!$ and $x(\log x - 1)$ for different values of x.

Table 14.11.1   Values of $\log x!$ and the values obtained by Stirling's approximation $x(\log x - 1)$ against different values of x.

x	$\log x!$	$x(\log x - 1)$
$1.0000 \times 10^{+02}$	$3.6374 \times 10^{+02}$	$3.6052 \times 10^{+02}$
$1.0000 \times 10^{+04}$	$8.2109 \times 10^{+04}$	$8.2103 \times 10^{+04}$
$1.0000 \times 10^{+06}$	$1.2816 \times 10^{+07}$	$1.2816 \times 10^{+07}$
$1.0000 \times 10^{+08}$	$1.7421 \times 10^{+09}$	$1.7421 \times 10^{+09}$

14.12 Summation and Integration

We show how numerical integration can be replaced by a summation and a correction term and *vice versa*. In Figure 14.12.1 we have plotted a

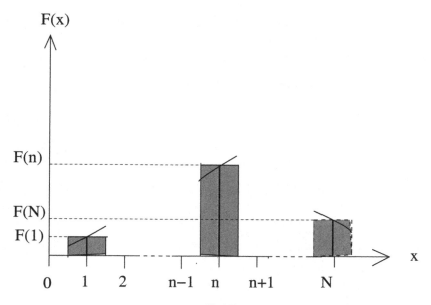

Fig. 14.12.1 Depiction of $\int_{1/2}^{N+1/2} F(x)dx$ and $\sum_{n=1}^{N} F(n)$.

function $F(x)$ in the range $n - \frac{1}{2} \leq x \leq n + \frac{1}{2}$ where n is an integer. In this range we make the approximation (Simpson's rule)

$$F(x) = a_n + b_n x + c_n x^2 \tag{14.12.1}$$

and get for the integral

$$
\int_{n-\frac{1}{2}}^{n+\frac{1}{2}} F(x)dx = \left[a_n x + \frac{b_n}{2}x^2 + \frac{c_n}{3}x^3 \right]_{n-\frac{1}{2}}^{n+\frac{1}{2}}
$$

$$
= a_n + b_n n + c_n n^2 + \frac{c_n}{12} = F(n) + \frac{c_n}{12}
$$

$$
= F(n) + \frac{F''(n)}{24}. \tag{14.12.2}
$$

Using $F''(n) = F'(n - \frac{1}{2}) - F'(n + \frac{1}{2})$, and taking summation over n from $n = 1$ to $n = N$, we obtain

$$\sum_{n=1}^{N} F(n) = \int_{\frac{1}{2}}^{N+\frac{1}{2}} F(x)dx + \frac{1}{24}\left[F'\left(\frac{1}{2}\right) - F'\left(N + \frac{1}{2}\right)\right]. \quad (14.12.3)$$

Shifting the origin along the abscissa we can write

$$\sum_{n=1}^{N} F(a + n) = \int_{a+\frac{1}{2}}^{a+N+\frac{1}{2}} F(x)dx + \frac{1}{24}\left[F'\left(a + \frac{1}{2}\right) - F'\left(a + N + \frac{1}{2}\right)\right].$$

$$(14.12.4)$$

This formula can be used both for evaluating a summation and also for calculating an integral.

14.13   Volume of an N-Dimensional Sphere

Theorem 14.13.1

$$V_N(R) = \frac{\pi^{N/2}R^N}{\Gamma(\frac{N}{2} + 1)}, \quad (14.13.1)$$

where the left hand side denotes the 'volume' of an

N − dimensional 'sphere' of 'radius' R.

Proof: The N-dimensional 'sphere' of 'radius' R is defined by the set $\{x_1, x_2, \cdots, x_N \mid x_1^2 + x_2^2 + \cdots + x_N^2 \leq R\}$.

For N = 1, obviously $V_1(R) = 2R$. Thus the theorem is true for N = 1.

Let the theorem be true for N = k, i.e. $V_k(R) = \frac{\pi^{k/2}R^k}{\Gamma(\frac{k}{2}+1)}$.

Considering a 'frustrum' of the (k+1)-dimensional 'sphere' of 'radius' R, shown in Figure 14.13.1, perpendicular to any polar axis and at a distance $z = R\cos\theta$ from the centre with 'thickness' $dz = R\sin\theta\, d\theta$ we get

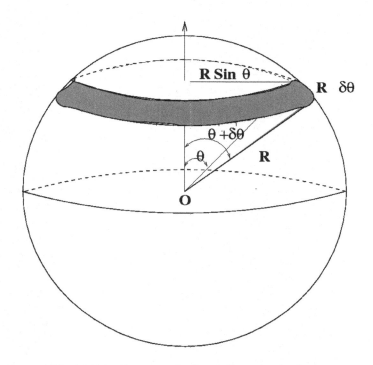

Fig. 14.13.1   Frustrum of a (k+1)-dimensional sphere.

$$V_{k+1}(R) = 2\int_0^{\frac{\pi}{2}} V_k(R\sin\theta)\,dz$$

$$= 2\int_0^{\frac{\pi}{2}} \frac{\pi^{\frac{k}{2}} R^k \sin^k\theta}{\Gamma\left(\frac{k}{2}+1\right)} R\sin\theta\,d\theta$$

$$= \frac{\pi^{\frac{k}{2}} R^{k+1}}{\Gamma\left(\frac{k}{2}+1\right)} 2\int_0^{\frac{\pi}{2}} \sin^{k+1}\theta\,d\theta$$

$$= \frac{\pi^{\frac{k}{2}} R^{k+1}}{\Gamma\left(\frac{k}{2}+1\right)} B\left(\frac{k}{2}+1, \frac{1}{2}\right)$$

$$= \frac{\pi^{\frac{k+1}{2}} R^{k+1}}{\Gamma\left(\frac{k+1}{2}+1\right)},$$

where use has been made of Equations 14.1.3, 14.1.5, and 14.1.9

Since the Theorem is true for N=1, by the method of transfinite induction it is thus proved for all values of N.

Bibliography

G Baym: *Neutron Stars & the Properies of Matter at High Density*
(Nordita, Copenhagen, 1977)
R P Feynman: *Statistical Mechanics*
(Benjamin/Cummings, Reading, MA, 1979)
K Huang: *Statistical Mechanics* (Wiley, Hoboken, NJ, 1963)
L D Landau & E M Lifshitz: *Statistical Physics, Parts I & II*
(Pergamon, Oxford, 1980)
S-K Ma: *Modern Theory of Critical Phenomena*
(Benjamin/Cummings, Reading, MA, 1976)
S-K Ma: *Statistical Mechanics* (World Scientific, Singapore,1985)
A Pais: *Subtle is the Lord* (OUP, Oxford, 1997)
R K Pathria: *Statistical Mechanics* (Elsevier, Oxford, 1996)
C J Pethick & H Smith: *Bose-Einstein Condensation in Dilute Gases*
(Nordita, Copenhagen, 1997)
F Reif: *Fundamentals of Statistical and Thermal Physics*
(Mc-Graw Hill, Singapore, 1988)
E Schatzman: *Physics & Astrophysics* (CERN, Geneva, 1970)
H E Stanley: *Introduction to Critical Phenomena* (OUP, Oxford, 1971)
M Toda, R Kubo, N Saito & N Hashitsume: *Statistical Physics, Parts I & II*
(Springer, Berlin, 1978)

Index